SMARTPHONES FROM AN APPLIED RESEARCH PERSPECTIVE

Edited by **Nawaz Mohamudally**

Smartphones from an Applied Research Perspective
http://dx.doi.org/10.5772/65172
Edited by Nawaz Mohamudally

Contributors

Diana Bueno, Roberto Munoz, Jean-Louis Marty, Robert LeMoyne, Timothy Mastroianni, Francisco Liebana-Cabanillas, Louie Zamora, Shigeru Kashihara, Noriyuki Suzuki, Abraham Rodríguez-Mota, Ponciano Jorge Escamilla-Ambrosio, Moisés Salinas Rosales, Paolo Dabove, Andrea Andrea, Vincenzo Di Pietra, Feng Liu, JianXun Cui, Davy Janssens, Geert Wets, Mario Cools, Cumhur Sahin, Mehdia El Khaddar Ajana, Mohammed Boulmalf, Missoki Azanlédji Boume

Published by InTech

Janeza Trdine 9, 51000 Rijeka, Croatia

Publishing Process Manager Romina Skomersic
Technical Editor SPi Global
Cover InTech Design team

First published October, 2017
Printed in Croatia
Legal deposit, Croatia: National and University Library in Zagreb

Additional hard copies can be obtained from orders@intechopen.com

Smartphones from an Applied Research Perspective, Edited by Nawaz Mohamudally
p. cm.
Print ISBN 978-953-51-3589-0
Online ISBN 978-953-51-3590-6

For EU product safety concerns:
IN TECH d.o.o., Prolaz Marije Krucifikse Kozulić 3, 51000 Rijeka, Croatia,
info@intechopen.com or visit our website at intechopen.com.

Meet the editor

Dr. Nawaz Mohamudally graduated in telecommunications from the University of Science and Technology of Lille in France and is currently an associate professor and chairman of the Research Degrees Committee at the University of Technology, Mauritius, where he had been the head of the School of Software Engineering and Business Informatics and the School of Innovative Technologies and Engineering. He was the chairman of the Internet Management Committee, an advisory unit to the Mauritian authority from 2010 to 2014. His core research areas are mobile and pervasive computing and data science. He has successfully supervised 6 PhD students and published 150+ refereed research articles and papers, co-authored 1 book and authored 6 book chapters. Dr Mohamudally was awarded the "Best Professor in Industrial Systems Engineering" by Africa Education Leadership Award in 2015.

Contents

Preface

The smartphone can be defined as a handheld device with an intelligent operating system with multiple usages. Its advent in the last decades is undoubtedly one of major technological innovations that has paramount economic and social impacts. The smartphone is today present in all spheres of business and society. For instance, most of the ERP solutions and web applications are available in mobile format. Figures show that the number of subscriptions has almost surpassed that of the world population and the resource constraints often attributed as major limitations for computation and storage are no longer valid, namely, with the complementarity of the cloud.

Likewise, emerging technologies such as IoT and edge computing promote smartphones as reliable end devices in real-life use cases. As a matter of fact, smart city mobile apps are examples of the necessity of the citizens to be acquainted with the efficient use of smart apps. Mobile app companies have betted on innovative minds and the billion dollar global market. But far beyond mobile apps, the smartphone has taken a phenomenal dimension from its physical and sensor attributes. This is very much perceptible in the utilisation of smartphone in healthcare scenarios and medical instrumentation. The multidisciplinary nature of research and development involving smartphones is pushing away the frontiers of knowledge, hence the need to compile latest advancements in the field in a concise and structured manner as provided by the current book.

This book disseminates smartphone-centric advance research achievements from authors coming from different corners across the globe, namely, Turkey, Mexico, France, Togo, Spain, Japan, and Italy. There are ten chapters covering issues and challenges in mobile malware and security, disaster recovery, healthcare, technology acceptance, mobile payment, misuse of mobile phones, education and future networking like positioning techniques. Researchers, practitioners and postgraduate students planning for MPhil/PhD studies will find ample food for thought. Readers will surely enjoy the array of topics made technically affordable with deep discussions and analysis. These were the main motivations behind the whole process and quality assurance in the development and realisation of this project.

We would like to express our appreciations to all the chapters' contributors and their respective institutions. We feel indebted to the Publishing Process Manager, Ms. Romina Skomersic, who has accompanied us in this endeavour from the outset. We would like to give special thanks to the InTech technical team for ensuring the quality of the manuscripts. We are very much grateful to the InTech Commissioning Editor, Ms. Anja Filipovic, for the success of this collaboration.

Associate Professor Dr. Nawaz Mohamudally
University of Technology,
Port Louis, Mauritius

Chapter 1

Smartphone and Portable Media Device: A Novel Pathway toward the Diagnostic Characterization of Human Movement

Robert LeMoyne and Timothy Mastroianni

Additional information is available at the end of the chapter

http://dx.doi.org/10.5772/intechopen.69961

Abstract

The application of wearable and wireless systems offers the capacity to ameliorate considerable strain on medical resources. In particular the smartphone and portable media device for quantifying human movement characteristics offers the opportunity to evaluate patients in a homebound environment remote from clinical resources and post-processing. Trial data can be easily transmitted as an email attachment with wireless connectivity to the Internet. The utility of the smartphone and portable media device has been demonstrated for quantifying gait, tendon reflex response, movement disorder, and rehabilitation exercise. Further evolution and potential has been demonstrated through the integration of machine learning to provide classification accuracy for differentiating between disparate human movement scenarios. The role of the smartphone and portable media device for quantifying human movement characteristics is further elucidated.

Keywords: smartphone, portable media device, wearable, wireless, wireless accelerometer, wireless gyroscope, gait analysis, reflex response, patellar tendon reflex, Parkinson's disease tremor, essential tremor, Virtual Proprioception, machine learning, rehabilitation, therapy

1. Introduction

Wearable and wireless systems, such as smartphones and portable media devices, have been demonstrated as functional wireless accelerometer and gyroscope platforms for quantifying human movement endeavors, such as gait, tendon reflex response, movement disorder, and rehabilitation exercise. Further evolution of the technology capability has been demonstrated with the integration of machine learning for attaining classification accuracy [1–3]. The success of these applications derives from preliminary research, development, testing, and evaluation of wireless accelerometer nodes that are essentially wearable for similar scenarios [4].

Inertial sensors such as accelerometers were originally proposed for the quantification of human movement, however at the time of this perspective they were not sufficiently evolved. The technology evolution of accelerometer systems proceeded in tandem with disparate industries relative to the biomedical, rehabilitation, and health industry [4, 5]. Quantification of rehabilitation status can facilitate the modification of a therapy intervention, especially with the expert clinical acuity of a trained expert, however traditional quantification apparatus, such as gait analysis quantification equipment, is generally restricted to a clinical environment [6, 7]. The role of the accelerometer system steadily progressed with the biomedical, rehabilitation, and health community [8]. Eventually with the development of wireless technology supporting inertial sensors, such as accelerometers and gyroscopes, other techniques, such as tethering, became effectively outmoded [9].

The wireless accelerometer has been successfully demonstrated as a wearable system for the quantification of human movement [4]. Tandem operated wireless accelerometer systems have been successfully applied to the quantified evaluation of hemiplegic gait [10–12]. Other associated endeavors have demonstrated the role of wireless accelerometers for quantifying reflex response and even reflex latency [13–19]. Further testing and evaluation has revealed the utility of wireless accelerometers for quantifying movement disorder tremor, such as for Parkinson's disease [20, 21]. Further evolution of the capability of wearable and wireless applications is featured through the research, development, testing, and evaluation of smartphones and portable media devices as wireless accelerometer and gyroscope platforms [1–3, 22].

Ever since the origins of the smartphone, one of its well known features is the observation that the screen will shift orientation based on movement and position. This capability is due to its inertial sensor, originally consisting of an accelerometer and now also a gyroscope. With this feature noted LeMoyne and Mastroianni sought to utilize this inertial sensor package to characterize human movement, much like their previous application of wireless accelerometers for similar endeavors [1–3, 22].

2. Smartphone: wireless accelerometer platform for quantifying human movement

Following this requirement definition LeMoyne and Mastroianni successfully acquired of an appropriate software application to record the accelerometer signal data and convey the data package as an email attachment. This capability transformed the smartphone from a telecommunication utility to a wearable and wireless accelerometer sensor system capable of measuring human motion within the context of an assortment of scenarios. Given the compact nature of the smartphone, it can be easily mounted about an assortment of readily identifiable anatomical mounting positions. Preliminary research, development, testing, and evaluation of the smartphone emphasized the role of quantifying gait and Parkinson's disease [1–3, 23, 24].

During 2010 LeMoyne and Mastroianni applied the smartphone as a wearable and wireless gait analysis device. The smartphone was equipped with an application that enabled it to function as a wireless accelerometer platform. The recorded data package of the acceleration waveform was conveyed by wireless transmission to the Internet as an email attachment [23].

Another major observation of the capabilities of this application was based on the remote nature between the experimental site and post-processing resources. The gait experiment was conducted in the region of Pittsburgh, Pennsylvania. However, the post-processing resources were situated in the greater Los Angeles area. The implications of the research endeavor are that with a suitable software application the smartphone operating as a wireless accelerometer can quantifying human movement characteristics, such as gait, with post-processing resources situated anywhere in the world. Essentially the email resource symbolizes a functional semblance of a cloud computing resource [23].

For the scope of gait analysis the smartphone was secured about the lateral malleolus near the ankle joint by an elastic band. Each gait experiment recorded on the order of 10 s of steady state walking. Temporal and kinematic parameters, such as the stance to stance temporal disparity and stance to stance time averaged acceleration, were acquired in an accurate and consistent manner [23]. Further testing and evaluation of the smartphone as a wireless accelerometer platform for gait analysis were successfully demonstrated using alternative mounting positions, such as the lateral epicondyle of the femur and the lumbar-sacral aspect of the spine [25, 26].

During 2010 LeMoyne and Mastroianni also applied the smartphone as a wearable and wireless accelerometer for the quantified acquisition of Parkinson's disease tremor. Measuring Parkinson's disease at the convenience of a patient's homebound setting is of paramount significance, in order to provide optimal acuity for expert clinical medical therapy intervention. A smartphone could measure Parkinson's disease hand tremor through mounting to the dorsum of the hand. With the experiment conducted in Pittsburgh, Pennsylvania and post-processing resources situated trans-continentally in greater Los Angeles, the application again demonstrates the ability to remotely situate experimental and post-processing resources. Tremor characteristics were successfully quantified for a person with Parkinson's disease and contrasted to a non-Parkinson's subject [24].

The implications of the research, development, test, and evaluation demonstrated by LeMoyne and Mastroianni elucidate the broad power of wearable and wireless systems, such as the smartphone as a wireless accelerometer platform. The smartphone can be easily mounted to effectively any portion of the body that best defined the characteristics of the human movement feature under consideration, such as the near the ankle joint for gait or dorsum of the hand for Parkinson's disease tremor. From the experimental site, the recorded accelerometer signal data can be conveyed through wireless Internet connectivity to a post-processing resource anywhere in the world. In essence a subject can access the best clinical resources in the world from the convenience of a familiar homebound and autonomous setting. Further testing and evaluation of the smartphone as a wireless accelerometer pertained to the quantification of reflex response.

The reflexes of the lower limb are synergistically interrelated with the function of gait [19]. Therefore with the success of the smartphone as a wireless accelerometer platform for quantifying gait, the patellar tendon reflex response logically should also be a readily quantifiable aspect of human movement. Preliminary testing and evaluation the smartphone as a wireless accelerometer for quantifying reflex response pertained to manual stimulation of the patellar tendon reflex [27]. The accurate and consistent quantification of the patellar tendon reflex can be further facilitated through the application of a potential energy impact pendulum for evoking the tendon reflex response with a prescribed amount of energy that is also targeted to a specified aspect of the patellar tendon [14–19].

Using a remote and effectively rural area as an experimental site, LeMoyne and Mastroianni applied a smartphone as a wireless accelerometer platform in conjunction with a potential energy impact pendulum. The integration of these devices readily acquired a recording of the reflex response acceleration waveform for the patellar tendon. The trial data was conveyed by wireless transmission to the Internet as an email attachment. Remotely situated post-processing resources applied software automation to quickly determine the efficacy of the experimental trial data. The findings advocate that the patellar tendon reflex response can be readily acquired through the integral application of the potential energy impact pendulum with the smartphone as a wireless accelerometer platform in an accurate and consistent manner [28]. Further establishment of the smartphone for quantifying the reflex response was demonstrated though the application of an artificial reflex system [29].

Further investigation of the opportunities for gait quantification emphasized the evaluation of gait for people with transtibial amputation. In consideration of people with transtibial amputation, they require a different mounting technique as opposed to merely applying an elastic band to secure the smartphone. In order to resolve this matter, a 3D printed mounting adapter was applied to secure the smartphone to the transtibial prosthesis. The smartphone conveniently measured the acceleration waveform of the subject's gait respective of the transtibial prosthesis. Automated software facilitated the post-processing endeavor with gait characteristics acquired in an accurate and reliable manner [30].

3. Smartphone: wireless gyroscope platform for quantifying human movement

With the continuous evolution of the smartphone eventually the gyroscope was integrated into the inertial sensor package. The gyroscope measures the rate of angular rotation, which offers a readily clinically identifiable metric, especially in consideration of the orientation of a particular joint [1–3, 22]. In particular the smartphone as a wireless gyroscope platform can quantify the characteristics of the patellar tendon reflex response [31].

In order to accurately and consistently quantify the reflex response of the patellar tendon through the application of a smartphone as a wireless gyroscope platform, a means of evoking the reflex with a similar level of accuracy and consistency is imperative. The potential energy impact pendulum attached to a reflex hammer was selected, which has been successfully researched, developed, tested, and evaluated, while demonstrating the ability to elicit the tendon reflex at prescribed levels of potential energy with predetermined targeting [14–19]. The integration of the potential energy impact pendulum and smartphone as a wireless gyroscope platform provide considerable opportunity for the accurate and consistent measurement of the reflex response.

Upon the experimentation trial data was conveyed by wireless transmission to the Internet as an email attachment. Each gyroscope signal was post-processed through a software automation program for the acquisition of a pertinent parameter, such as the maximum gyroscope signal of the response. This preliminary investigation of the application of the smartphone as a wireless gyroscope platform in conjunction with a potential energy impact pendulum demonstrated the ability to accurately and reliability quantify the patellar tendon reflex response in terms of the gyroscope signal [31].

Further testing and evaluation of the smartphone as a wireless gyroscope platform investigated the capability to quantify gait. Using a mounting strategy that positioned the smartphone about the trunk, preliminary gyroscope signal data was acquired for gait. The findings implied that the smartphone as a wireless gyroscope platform could also be applied for identifying gait characteristics [32].

Later in the chapter the expanded role of the smartphone as a wireless gyroscope platform is demonstrated for the domain of machine learning classification. An assortment of machine learning algorithms are applied, for which the gyroscope signal provides a suitable feature set. Machine learning using the smartphone as a wireless gyroscope emphasizes the domains of classifying therapy exercise status and differentiating hemiplegic affected and unaffected reflex pairs.

The portable media device represents a similar wireless sensor platform, first with the accelerometer and later with the gyroscope. The portable media device and smartphone are supported by the same software application, such as for the iPod and iPhone. The appropriateness of either device for an experiment quantifying human movement is at the discretion of the research team. For example, the portable media device has a reduced telecommunications footprint, which is restricted to local wireless connectivity for access to the Internet; however the portable media device in general is somewhat lighter and relatively cheaper [1–3]. The following section investigates the capability of the portable media device of quantifying human movement, such as reflex and gait, first from the perspective of a wireless accelerometer platform.

4. Portable media device: wireless accelerometer platform for quantifying human movement

Preliminary testing and evaluation of the portable media device pertained to the evaluation of gait. The portable media device was mounted proximal to the ankle joint and secured by an elastic band. Connectivity to the Internet was enabled through local wireless connectivity, for which the trial data samples were conveyed as email attachments to a predetermined address.

Because of the opportunities facilitated by Internet connectivity though the email address, the experimental site and post-processing resources are trans-continentally situated on either side of the United States of America [33].

Upon testing and evaluation notable advantages of the portable media device are evident. The portable media device is lighter than the smartphone. Because it only relies on local wireless Internet connectivity, there is only a fixed cost for the device as opposed to a continuous marginal cost to sustain a telecommunication footprint. The portable media device demonstrated the ability to acquire gait data in an accurate and consistent manner [33]. Further testing and evaluation of the portable media device as wireless accelerometer gait analysis system about the lower aspect of the trunk successfully identified temporal features of gait [34].

As possession of two portable media devices is relatively more feasible than ownership of two smartphones, the application of tandem mounted portable media devices to quantify the disparity present in hemiplegic gait can be readily accomplished. Both portable media devices functioning as wearable and wireless accelerometers were mounted proximal to the lateral malleolus near the ankle joint for the hemiplegic affected leg and unaffected leg. Again the experimental and post-processing resources were trans-continentally situated with the trial data wirelessly conveyed to the Internet as an email attachment [35].

The post-processing aspect derived the affected leg and unaffected leg temporal disparity of stance to stance, time averaged acceleration of stance to stance, and the affected leg/unaffected leg ratio for stance to stance time averaged acceleration less the offset. These quantified gait parameters demonstrated considerable accuracy, consistency, and reliability. From an inferential statistical perspective the temporal disparity of stance to stance and time averaged acceleration of stance to stance for the affected leg contrasted to the unaffected leg demonstrated statistical significance. Future evolutions to the concept envision automated derivation of parameters and machine learning classification [35].

As previously disclosed gait and reflex of the lower limb, such as the patellar tendon reflex, are neurologically associated [19, 36]. Therefore the portable media device provides a useful means of quantifying the patellar tendon reflex response. In particular the localized wireless connectivity to the Internet is interrelated with the generally indoor and localized environment inherent for obtaining a series of patellar tendon reflex response samples. LeMoyne and Mastroianni demonstrated in 2011 the ability to accurately and consistently characterize the reflex response through manual supramaximal stimulation using a portable media device as a wireless accelerometer platform [37].

Further application of the portable media device as a wireless accelerometer platform amalgamated the utility of the potential energy impact pendulum. The portable media device was mounted proximal to the lateral malleolus of the ankle joint by an elastic sock. Each trial sample was transmitted to the Internet through localized wireless connectivity. Post-processing consolidated the accelerometer signals in three dimensions to a single signal presenting acceleration magnitude. The maximum acceleration of the reflex response was the parameter of interest. The application for quantifying reflexes demonstrated the capacity to characterize the reflex response in a considerably accurate and consistent manner [38].

5. Portable media device: wireless gyroscope platform for quantifying human movement

Further evaluation of the portable media device and its associated software for the capability as a wireless gyroscope platform developed. In particular since the reflex response of the patellar tendon is jointed about the knee, therefore the gyroscope signal revealing the rate of angular rotation provides a highly clinically relevant signal. The trial data of the portable media device functioning as wireless gyroscope platform is wirelessly conveyed to the Internet as an email attachment [31, 39]. Its implications are that the post-processing location, such as in America, can be totally remote from the experimental site, such as remote Tibet.

While on travel LeMoyne and Mastroianni decided to truly test the robust capability of the portable media device as a wireless gyroscope platform for quantifying the patellar tendon reflex response with the experimental site selected as Lhasa, Tibet and the subsequent post-processing in Flagstaff, Arizona in the United States of America. An advantage of the portable media device is the sole need to locally connect to a local wireless Internet zone. The data samples were obtained through supramaximal stimulation of the patellar tendon reflex. The portable media device was mounted to the lateral malleolus of the ankle through a sock. The data was stored at an assigned email site for later post-processing, for which the maximum of the gyroscope signal characterizing the reflex response was acquired with considerable accuracy and consistency [39].

6. Machine learning classification with the smartphone and portable media device as wireless accelerometer and gyroscope platforms

The synergy of devices, such as smartphones and portable media devices, for quantifying human movement and machine learning offers the exciting opportunity of highly objective assessment, automated computer aided diagnosis, and prognostic forecasting. LeMoyne and Mastroianni have recently emphasized the classification of movement features of the hemiplegic affected and unaffected side. Preliminary test and evaluation has demonstrated considerable classification accuracy [1, 2, 22].

These accomplishments have featured the use of Waikato Environment for Knowledge Analysis (WEKA). WEKA is equipped with a considerable array of machine learning algorithms, such as the support vector machine, multilayer perceptron neural network, J-48 decision tree, logistic regression, and K-nearest neighbors. The appropriate machine learning algorithm is selected at the discretion of the research team with consideration to the classification endeavor at hand. The feature set that comprises the data is organized into an Attribute-Relation File Format (ARFF) file. The ARFF file is derived primarily from a series of numeric attributes that appropriately quantify the classes to be distinguished [40–42].

The ability for even a skilled team of clinicians to conclusively decide upon the presence of a hemiplegic reflex pair is a subject of contention [19, 36]. Considerable classification accuracy was attained through the amalgamation of a portable media device as a wireless accelerometer platform, a potential energy impact pendulum, and a machine learning algorithm. Each reflex response sample of the affected hemiplegic reflex response and unaffected reflex response was transmitted through local wireless connectivity as email attachments for post-processing. Software automation consolidated the trial data into a cohesive feature set. The selected machine learning algorithm was the support vector machine. This research demonstrates the considerable promise of the integration of smartphones and portable media devices as wearable and wireless inertial sensor platforms in combination with machine learning algorithms for classification of perceptively distinguishable human movement features [43].

Further investigation regarding the utility of the portable media device as a wireless accelerometer platform extended to the domain of assistive devices for gait, such as a cane. An issue with using a cane involves the concern that the patient is properly using the cane per the advice of a skilled therapist. However, such instruction is only realistically feasible respective of a relatively brief clinical appointment. A portable media device functioning as a wireless accelerometer platform can monitor the usage of a cane while the subject is walking, and the data package can be conveyed by local wireless connectivity to the Internet as an email attachment. Considerable classification accuracy was attained using logistic regression as a machine learning algorithm to distinguish between correct and incorrect cane usage scenarios with the recorded accelerometer signal deriving the feature set [44].

LeMoyne and Mastroianni demonstrated the potential of integrating a portable media device as a wireless gyroscope platform for quantifying the patellar tendon reflex response and machine learning for the classification of a hemiplegic reflex pair. The tendon reflex was elicited through supramaximal stimulation from a manually operated reflex hammer. The portable media device functioning as a wireless gyroscope platform was mounted proximal to the lateral malleolus as demonstrated in **Figure 1** [45].

In order to consolidate the affected leg and unaffected leg reflex response into a feature set their respective gyroscope signals needed to be considered, such as presented in **Figure 2**. Three numeric attributes comprehensively characterize the nature of the reflex response:

- maximum angular rate of rotation;
- minimum angular rate of rotation;
- time disparity between maximum and minimum angular rate of rotation [45].

These three numeric attributes represent the reflex response in terms of both kinematic and temporal features. The feature set was evaluated through WEKA using the J48 decision tree. An advantage of the WEKA J48 decision tree is the ability to visualize the decision tree, which may infer the most predominant aspects of the feature set [45].

The J48 decision tree is illustrated in **Figure 3**. Note that the feature set attribute TimeMaxMinGamma is clearly the dominant attribute for distinguishing between the hemiplegic affected leg patellar tendon reflex response and the unaffected leg patellar tendon reflex response. The research findings demonstrate the capacity of machine learning, such as the J48 decision tree, for attaining considerable classification accuracy that distinguishes between a hemiplegic affected patellar tendon

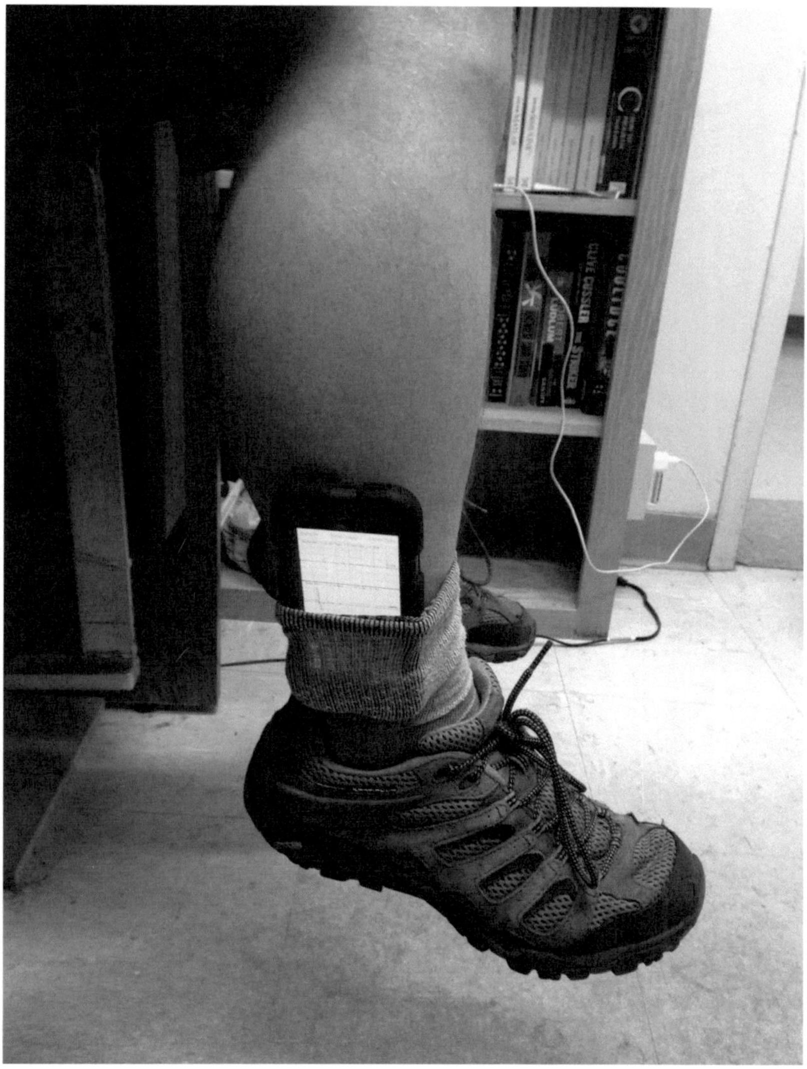

Figure 1. Portable media device as a wireless gyroscope platform for quantifying the patellar tendon reflex response [45].

reflex response and correlated unaffected patellar tendon reflex response through the quantification of a portable media device functioning as a wireless gyroscope platform [45].

An extension of the utility of using the portable media device as a wireless gyroscope platform with machine learning was applied to classify the usage of a proposed rehabilitation system. An ankle rehabilitation system intended to promote dorsiflexion was developed through 3D printing. The operation of the ankle rehabilitation system was quantified by mounting a smartphone functioning as wireless gyroscope platform to the foot plate of the ankle rehabilitation system [46].

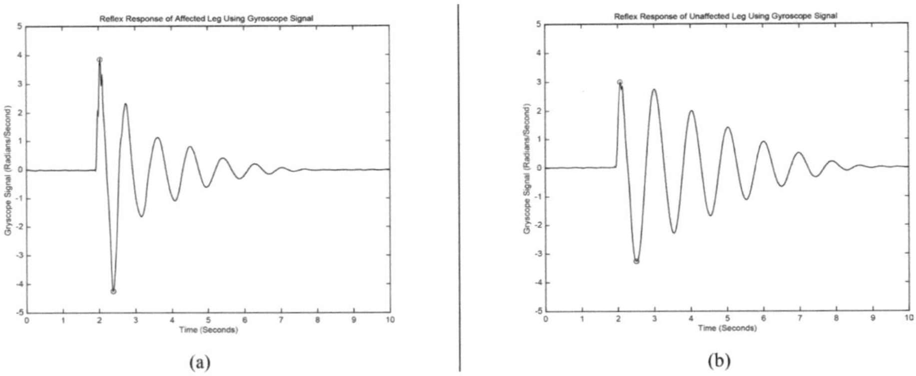

(a) (b)

Figure 2. Gyroscope signals for a hemiplegic reflex pair (affected leg (a) and unaffected leg (b)) for the patellar tendon reflex response [45].

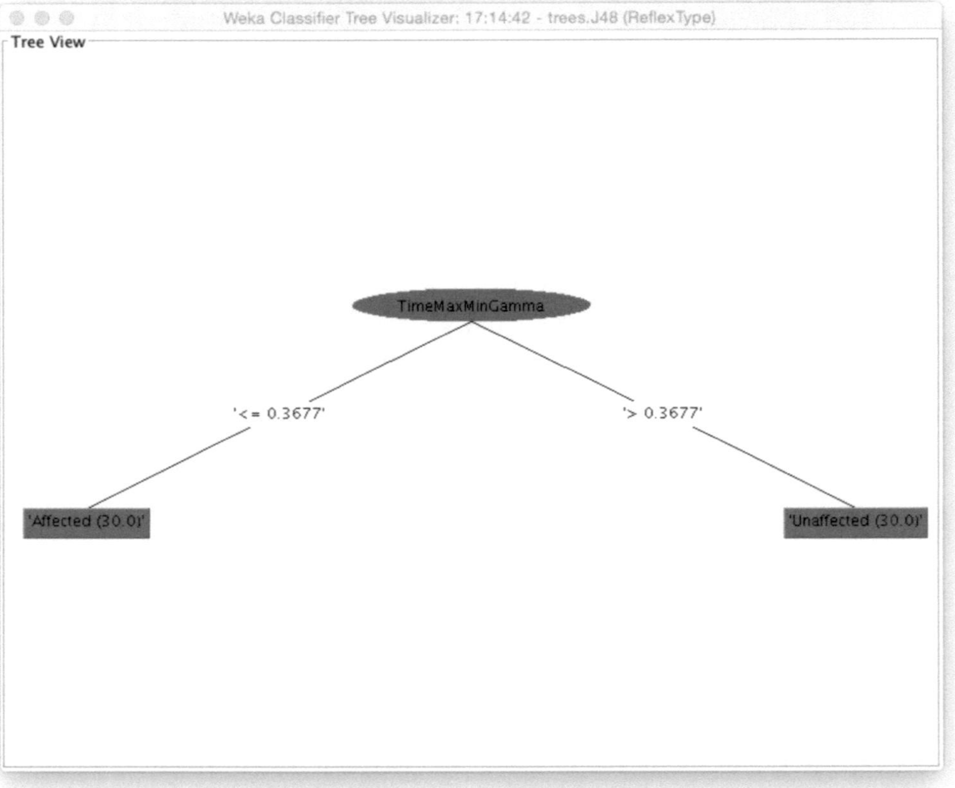

Figure 3. J48 decision tree applied for attaining considerable classification accuracy for differentiating between a hemiplegic reflex pair for the patellar tendon reflex response [45].

The ability of the smartphone functioning as wireless gyroscope platform to convey its data sample as an email attachment through Internet connectivity makes the integral application suitable for homebound therapy scenarios. A subject could conduct a therapy session at home, and then the post-processing with machine learning classification could provide a therapist with advanced acuity of subject rehabilitation status. The gyroscope signal was distilled into a feature set for machine learning classification through a support vector machine, for which considerable classification accuracy was achieved [46].

Deep brain stimulation has been demonstrated as an efficacious treatment strategy for people with movement disorders that are intractable with medical therapy intervention. In particular essential tremor can be treated through the application of deep brain stimulation. However an issue is evident with respect to the considerable array of tuning parameters, which can present a daunting task for even an expert clinician. A smartphone as a wireless accelerometer platform presents a promising means for quantifying the efficacy of deep brain stimulation for people with essential tremor [47, 48].

Preliminary research, development, testing, and evaluation of the capability to a smartphone as a wireless accelerometer platform to facilitate the tuning of a deep brain stimulator for ameliorating symptoms of essential tremor pertained to attaining machine learning classification accuracy for differentiating between the deep brain stimulator in 'On' and 'Off' states. Regarding both of these states a subject with essential tremor was tasked with reaching and grasping a lightweight object with a smartphone mounted to the dorsum of the hand. The trial data of the smartphone functioning as wireless accelerometer platform transmitted the recording as an email attachment with connectivity to the Internet. Subsequent post-processing of the acceleration signals derived a feature set that was applied to a machine learning algorithm. The most appropriate algorithm for the classification task under consideration based on the expertise of the research team was the support vector machine. The support vector machine achieved considerable classification accuracy for differentiating between a person with essential tremor conducting a simple reaching and grasping task with deep brain stimulation in 'On' and 'Off' status based on the acceleration signal of a smartphone functioning as a wireless accelerometer platform [47].

Reduced arm swing is a subject of concern for people with hemiplegic gait. Objectively quantifying the severity of reduced arm swing for the affected side contrasted to the unaffected side may enable a therapy intervention to ameliorate the impact of reduced arm swing during gait. Given the inherently rotational nature of arm swing during gait, this scenario is particularly relevant to the application of a smartphone functioning as a wireless gyroscope platform [49].

Especially regarding an outdoor environment for walking and evaluating reduced arm swing respective of hemiplegic gait, the smartphone functioning as a wireless gyroscope platform can readily transmit experimental data as an email attachment through its broad wireless connectivity to the Internet. The smartphone can be easily mounted proximal to the wrist joint through an armband as illustrated in **Figure 4**. Post-processing can be conveniently conducted in a remote location [49].

Figure 4. Mounting of the smartphone as a wireless gyroscope platform for quantifying reduced arm swing [49].

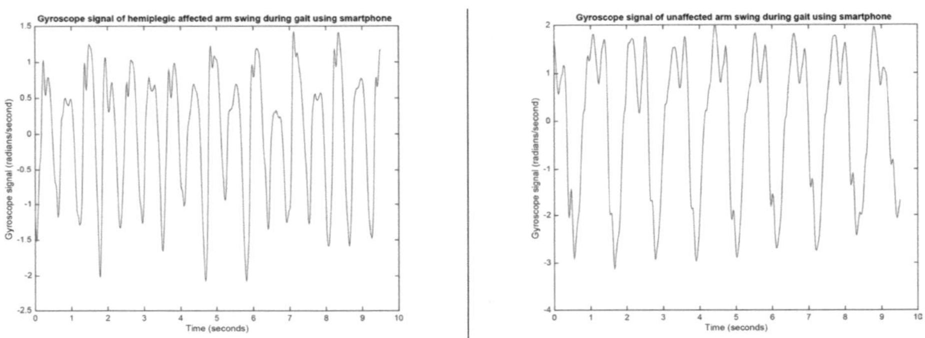

(a) (b)

Figure 5. Gyroscope signals for reduced arm swing regarding the hemiplegic affected arm (a) and unaffected arm (b) [49].

Figure 5 demonstrates the sample gyroscope signals of reduced arm swing of the affected arm contrasted to the unaffected arm. Based on the consideration of the gyroscope signal data the feature set was composed of aspects, such as descriptive statistics. A multilayer perceptron neural network was selected as the machine learning algorithm as provided in **Figure 6**. Considerable classification accuracy was attained for differentiating reduced arm swing regarding the hemiplegic affected arm and unaffected arm [49]. Similar results were also achieved with respect to another form of reduced arm swing manifested by Erb's Palsy [50].

Further refinement of the machine learning classification of a hemiplegic patellar tendon reflex pair was enabled through the application of the potential energy impact pendulum that provides predetermined amounts of potential energy to evoke the reflex and targeting of the reflex hammer. Upon observation of the obtained gyroscope signals the feature set was composed of three attributes:

http://dx.doi.org/10.5772/intechopen.69961

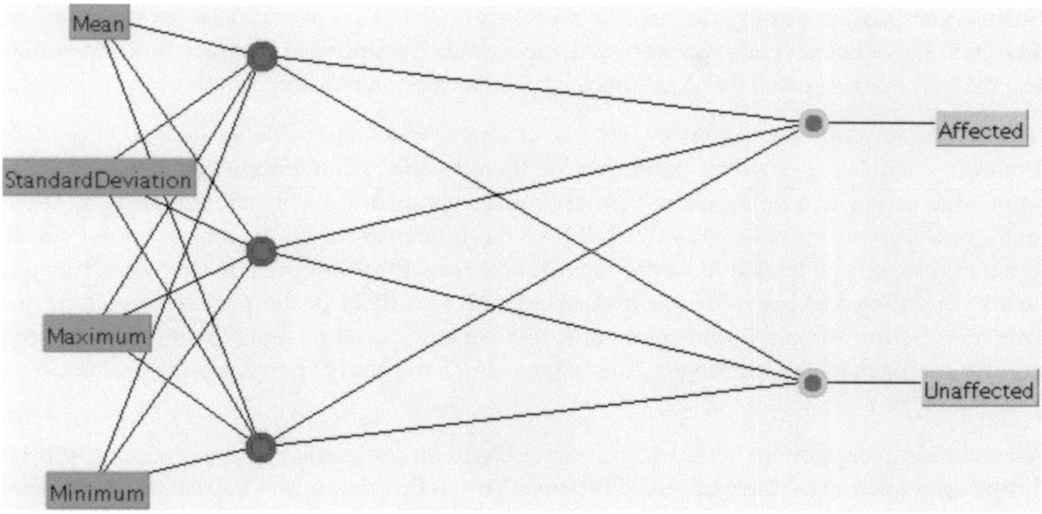

Figure 6. Multilayer perceptron neural network for classifying reduced arm swing regarding the hemiplegic affected arm and unaffected arm [49].

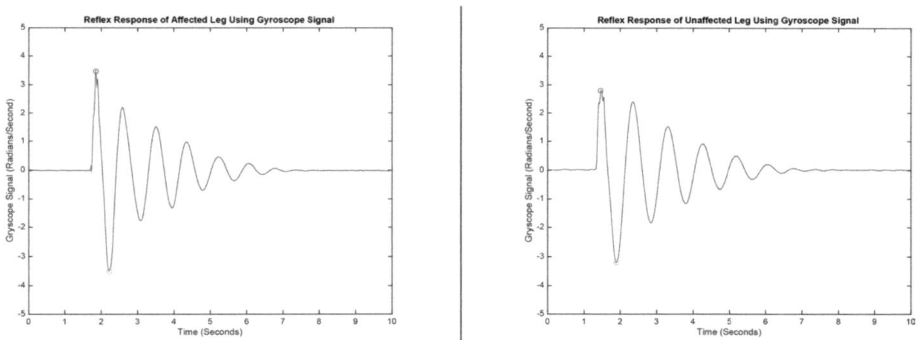

(a) (b)

Figure 7. Gyroscope signals for the affected leg (a) and unaffected leg (b) regarding the patellar tendon reflex response [52].

- maximum angular rate of rotation;
- minimum angular rate of rotation;
- time disparity between maximum and minimum angular rate of rotation.

The multilayer perceptron neural network was selected as the machine learning algorithm, for which considerable classification accuracy was attained distinguishing between a hemiplegic patellar tendon reflex response and its associated unaffected patellar tendon reflex response [51].

A similar experiment was applied using supramaximal stimulation of the patellar tendon reflex through a manually operated reflex hammer. **Figure 7** illustrates the acquired gyroscope signals for the affected leg and unaffected leg regarding the patellar tendon reflex response.

Subsequent post-processing derived the multilayer perceptron neural network presented in **Figure 8**. This research endeavor achieved appreciable classification accuracy for differentiating the hemiplegic (affected and unaffected) patellar tendon reflex pair [52].

Two recent developments regarding the role of smartphones and portable media device functioning as wireless gyroscope platforms for therapy and rehabilitation exercise have been applied to eccentric training using Virtual Proprioception and wobble board therapy. These two applications emphasize the flexibility of these devices, since the smartphone uses its broader telecommunication footprint regarding Virtual Proprioception for eccentric training in a gym setting and the more localized wireless connectivity of the portable media device in a homebound setting. Both experiments involve automated post-processing to develop a feature set for machine learning classification using a multilayer perceptron neural network [53, 54].

Virtual Proprioception for eccentric training is based on the successful application of Virtual Proprioception for real time gait rehabilitation. Virtual Proprioception for real time gait rehabilitation applied biofeedback based on a wireless accelerometer signals from a hemiplegic affected leg and unaffected leg while walking. Both auditory and visual feedback enabled the subject to modify gait in real time to achieve a closer state of parity regarding the acceleration waveforms of both legs. The wireless inertial sensors, such as the accelerometer, provide a virtual alternative to the neurological basis of proprioception [55, 56].

Eccentric strength training has be proposed as an highly effective means of strength training, however the rate of change is significant for the quality of the eccentric training event. Visual real time feedback from the gyroscope signal of a smartphone functioning as wireless gyroscope platform enables Virtual Proprioception for eccentric training. **Figure 9** illustrated the mounting of the smartphone through an armband proximal to the wrist joint for a biceps curl using Virtual Proprioception for eccentric training [54].

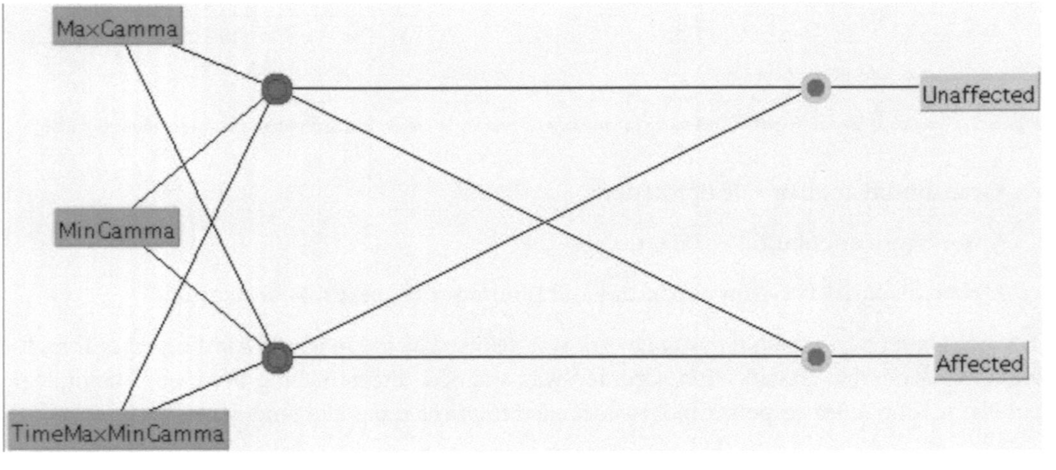

Figure 8. Multilayer perceptron neural network for classifying the affected leg and unaffected leg regarding the patellar tendon reflex response [52].

Figure 9. Virtual Proprioception for eccentric training demonstrated with a smartphone functioning as a wireless gyroscope platform secured proximal to the wrist through an armband [54].

During the eccentric muscle lengthening phase of a biceps curl the subject while using Virtual Proprioception for eccentric training uses the visual feedback from the smartphone gyroscope signal to ensure that a prescribed threshold is not exceeded. This inertial sensor acuity provides a virtual and quantified representation of neurologically derived proprioception. The control aspect of the experiment consists of an eccentric phase of a biceps curl without Virtual Proprioception. As illustrated in **Figure 10** these gyroscope signals display visually notable disparity [54].

The experimental trial data was transmitted wirelessly to the Internet as an email attachment. Automated post-processing of the data package developed a feature set based on the

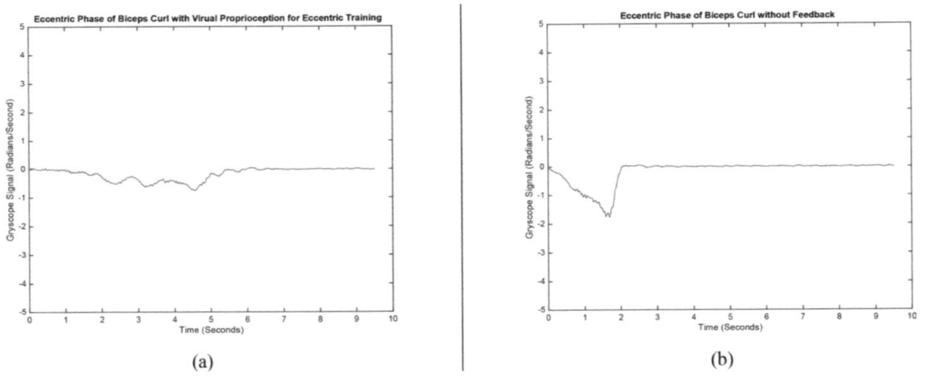

Figure 10. Gyroscope signals for eccentric phase of a biceps curl respective of Virtual Proprioception for eccentric training (a) and without feedback (b) [54].

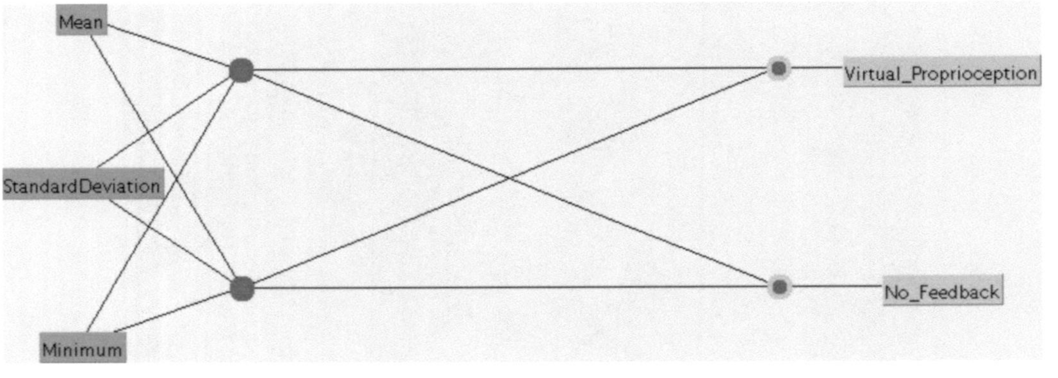

Figure 11. Multilayer perceptron neural network applied for classifying between scenarios of eccentric phase for a set of biceps curls respective of Virtual Proprioception for eccentric training and without feedback through Virtual Proprioception [54].

gyroscope signal data. Through the application of a multilayer perceptron neural network illustrated in **Figure 11** considerable classification accuracy was achieved for distinguishing between scenarios with Virtual Proprioception for eccentric training and eccentric training without feedback through Virtual Proprioception [54].

A wobble board is a therapy device that promotes rehabilitation of the ankle foot complex. A portable media device as shown in **Figure 12** can be readily mounted to the wobble board to provide advanced acuity as a wireless gyroscope platform. The rotation of the wobble board is notably disparate regarding the hemiplegic affected ankle and unaffected ankle upon observation. The observed disparity can be quantified through the wireless gyroscope platform from the portable media device and conveyed as an email attachment through wireless connectivity to the Internet. Post-processing consolidates the gyroscope signal into a feature set. With a multilayer perceptron neural network considerable classification accuracy was achieved for differentiating between the hemiplegic affected ankle and unaffected ankle [53].

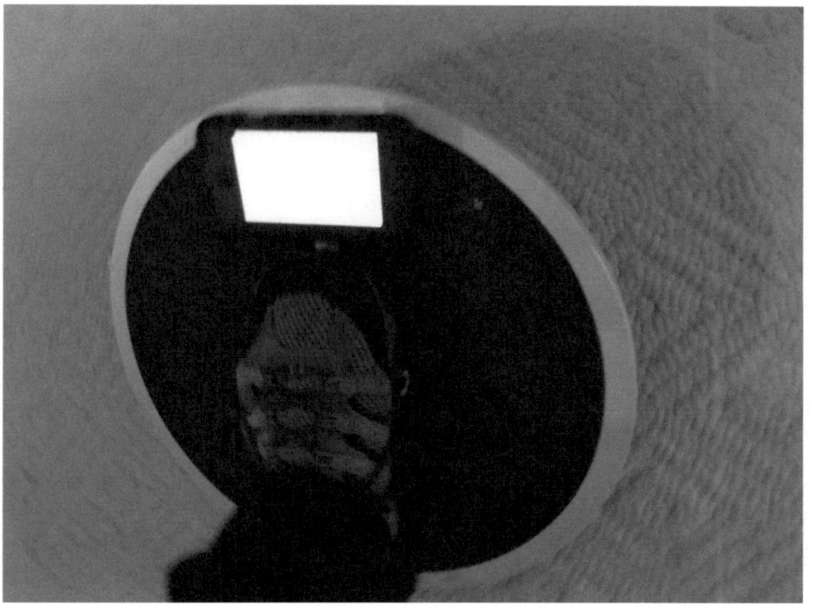

Figure 12. Wobble board for therapy of the ankle foot complex with a portable media device functioning as a wireless gyroscope platform [53].

7. Future concepts for smartphones and portable media devices, sensor fusion

In the near future the application of sensor fusion will likely be achieved broadly throughout the domain of smartphones and portable media devices as wearable and wireless accelerometer and gyroscope sensors. Sensor fusion involves the tandem operation of multiple inertial sensors, such as an accelerometer and gyroscope. With the combined signal of both accelerometer and gyroscope signal the actual spatial location in terms of displacement, velocity, and acceleration of the sensor mounting position can be determined [57–59]. For example, the spatial representation of foot displacement can be ascertained during gait [57].

Traditional sensor fusion however requires a considerable sampling rate that exceeds the feasible sampling threshold for smartphones and portable media devices as wearable and wireless systems [1–3, 57]. Post-processing resources are generally not portable, and techniques, such as the Kalman filter are required to serve as an orientation filter [57–59]. An orientation filter that satisfies the computational bounds associated with smartphones and portable media devices would be desirable for sensor fusion applications.

Madgwick has proposed a gradient descent algorithm as an orientation filter that is feasible for smartphones and portable media devices for the scope of sensor fusion. The IMU version requires only on the order of roughly 100 computational operations per filter update. The filter is also amenable to reduced sampling rates, such as 10 Hz [58, 59]. These capabilities make the gradient descent algorithm a realistic orientation filter with regards to sensor fusion for smartphones and portable media devices.

Sensor fusion also requires a novel application for smartphones and portable media devices that can simultaneously record the accelerometer and gyroscope signal for the same sample. Cognition Engineering has recently developed a smartphone and portable media device application suitable for acquiring both the accelerometer and gyroscope signal in a simultaneous manner. An example of the Cognition Engineering application has been demonstrated with regards to a person with essential tremor conducting a reach and grasp task with a deep brain stimulator in 'On' and 'Off' mode [60].

8. Network centric therapy

These trends evidenced through the development of smartphones and portable media devices as wireless accelerometer and gyroscope platforms that are effectively wearable advocate the development of Network Centric Therapy. With the development of Network Centric Therapy a patient and therapist could reside hundreds or thousands of mile remote. Rather than scheduling a traditional clinical appointment, therapy exercises and evaluation could be measured and quantified by systems, such as smartphones and portable media devices as wireless accelerometer and gyroscope platforms, for the quantification of human movement. The patient in a familiar setting of choice could conduct each therapy exercise and evaluation, and the acquired data could be transmitted wirelessly through Internet connectivity as email attachments. The post-processing could apply machine learning for augmented acuity for the therapist to determine critical transition phases of the therapy prescription and optimization of the rehabilitation experience.

9. Conclusion

Smartphones and portable media devices have been demonstrated through progressive research, development, testing, and evaluation as wearable and wireless accelerometer and gyroscope platforms for the quantification of human movement. In particular their utility has been advocated in domains, such as quantifying gait, tendon reflex response, movement disorder, and rehabilitation exercise. Experimental data can be readily transmitted through wireless connectivity to the Internet as an email attachment. Post-processing resources and the experimental location can be remotely situated anywhere in the world. Post-processing the inertial sensor signal data into a feature set enables machine learning classification. Research has demonstrated the capacity of an assortment of machine learning algorithms to achieve considerable classification accuracy for differentiating disparate human movement scenarios, such as contrasting a hemiplegic affected side to its associated unaffected side. Smartphones and portable media devices functioning as wireless accelerometer and gyroscope platforms are envisioned to facilitate the development of Network Centric Therapy, which is predicted to radically advance the therapy and rehabilitation experience.

Author details

Robert LeMoyne[1]* and Timothy Mastroianni[2]

*Address all correspondence to: rlemoyne07@gmail.com

1 Department of Biological Sciences and Center for Bioengineering Innovation, Northern Arizona University, Flagstaff, Arizona, USA

2 Independent, Pittsburgh, PA, USA

References

[1] LeMoyne R, Mastroianni T. Use of smartphones and portable media devices for quantifying human movement characteristics of gait, tendon reflex response, and Parkinson's disease hand tremor. Mobile Health Technologies: Methods and Protocols. 2015:335-358

[2] LeMoyne R, Mastroianni T. Wearable and wireless gait analysis platforms: Smartphones and portable media devices. Wireless MEMS Networks and Applications. 2017:129-152

[3] LeMoyne R, Mastroianni T. Telemedicine perspectives for wearable and wireless applications serving the domain of neurorehabilitation and movement disorder treatment. Telemedicine. 2016:1-10

[4] LeMoyne R, Coroian C, Cozza M, Opalinski P, Mastroianni T, Grundfest W. The merits of artificial proprioception, with applications in biofeedback gait rehabilitation concepts and movement disorder characterization. Biomedical Engineering. 2009:165-198

[5] Culhane KM, O'connor M, Lyons D, Lyons GM. Accelerometers in rehabilitation medicine for older adults. Age and Ageing. 2005 Nov 1;**34**(6):556-560

[6] Dobkin BH. The Clinical Science of Neurologic Rehabilitation; Oxford University Press, New York, 2003

[7] Perry J. Gait Analysis: Normal and Pathological Function; Slack, Thorofare, 1992

[8] LeMoyne R, Coroian C, Mastroianni T, Grundfest W. Accelerometers for quantification of gait and movement disorders: A perspective review. Journal of Mechanics in Medicine and Biology. 2008 Jun;**8**(02):137-152

[9] Patel S, Park H, Bonato P, Chan L, Rodgers M. A review of wearable sensors and systems with application in rehabilitation. Journal of Neuroengineering and Rehabilitation. 2012 Apr 20;**9**(1):21

[10] LeMoyne R, Coroian C, Mastroianni T, Grundfest W. Wireless accelerometer assessment of gait for quantified disparity of hemiparetic locomotion. Journal of Mechanics in Medicine and Biology. 2009 Sep;**9**(03):329-343

[11] LeMoyne R, Coroian C, Mastroianni T. Wireless accelerometer system for quantifying gait. In: ICME International Conference on 2009 Apr 9 Complex Medical Engineering (CME): IEEE; 2009. pp. 1-4

[12] LeMoyne R, Mastroianni T, Grundfest W. Wireless accelerometer system for quantifying disparity of hemiplegic gait using the frequency domain. Journal of Mechanics in Medicine and Biology. 2013 Jun;**13**(03):1350035

[13] LeMoyne R, Jafari R, Jea D. Fully quantified evaluation of myotatic stretch reflex. In: 35th Society for Neuroscience Annual Meeting; Washington DC; 2005 Nov 12

[14] LeMoyne R, Dabiri F, Jafari R. Quantified deep tendon reflex device, second generation. Journal of Mechanics in Medicine and Biology. 2008 Mar;**8**(01):75-85

[15] LeMoyne R, Coroian C, Mastroianni T, Grundfest W. Quantified deep tendon reflex device for response and latency, third generation. Journal of Mechanics in Medicine and Biology. 2008 Dec;**8**(04):491-506

[16] LeMoyne R, Mastroianni T, Kale H, Luna J, Stewart J, Elliot S, Bryan F, Coroian C, Grundfest W. Fourth generation wireless reflex quantification system for acquiring tendon reflex response and latency. Journal of Mechanics in Medicine and Biology. 2011 Mar;**11**(01):31-54

[17] LeMoyne R, Mastroianni T, Coroian C, Grundfest W. Wireless three dimensional accelerometer reflex quantification device with artificial reflex system. Journal of Mechanics in Medicine and Biology. 2010 Sep;**10**(03):401-415

[18] LeMoyne R, Coroian C, Mastroianni T. Wireless accelerometer reflex quantification system characterizing response and latency. In: Annual International Conference of the IEEE Engineering in Medicine and Biology Society, 2009 (EMBC 2009): IEEE; 2009 Sep 3. pp. 5283-5286

[19] LeMoyne RC. Wireless Quantified Reflex Device [Ph.D. Dissertation]. University of California, Los Angeles; 2010

[20] LeMoyne R, Coroian C, Mastroianni T. Quantification of Parkinson's disease characteristics using wireless accelerometers. In: ICME International Conference on 2009 Apr 9 Complex Medical Engineering (CME): IEEE; 2009. pp. 1-5

[21] LeMoyne R, Mastroianni T, Grundfest W. Wireless accelerometer configuration for monitoring Parkinson's disease hand tremor. Advances in Parkinson's Disease. 2013 May 10;**2**(02):62-67

[22] LeMoyne R. Testing and evaluation strategies for the powered prosthesis, a global perspective. Advances for Prosthetic Technology. 2016:37-58

[23] LeMoyne R, Mastroianni T, Cozza M, Coroian C, Grundfest W. Implementation of an iPhone as a wireless accelerometer for quantifying gait characteristics. In: Annual International Conference of the IEEE 2010 Aug 31 Engineering in Medicine and Biology Society (EMBC): IEEE; 2010. pp. 3847-3851

[24] LeMoyne R, Mastroianni T, Cozza M, Coroian C, Grundfest W. Implementation of an iPhone for characterizing Parkinson's disease tremor through a wireless accelerometer application. In: Annual International Conference of the IEEE 2010 Aug 31 Engineering in Medicine and Biology Society (EMBC): IEEE; 2010. pp. 4954-4958

[25] LeMoyne R, Mastroianni T, Cozza M, Coroian C. iPhone wireless accelerometer application for acquiring quantified gait attributes. In: ASME 2010 5th Frontiers in Biomedical Devices Conference: American Society of Mechanical Engineers; 2010 Sep 20. pp. 19-20

[26] LeMoyne R, Mastroianni T, Cozza M, Coroian C. Quantification of gait characteristics through a functional iPhone wireless accelerometer application mounted to the spine. In: ASME 2010 5th Frontiers in Biomedical Devices Conference: American Society of Mechanical Engineers; 2010 Sep 20. pp. 87-88

[27] LeMoyne R, Coroian C, Mastroianni T, Cozza M, Grundfest W. Quantification of reflex response through an iPhone wireless accelerometer application. In: Proceedings of 40th Society for Neuroscience Annual Meeting; San Diego, CA; 2010 Nov 13

[28] LeMoyne R, Mastroianni T, Grundfest W, Nishikawa K. Implementation of an iPhone wireless accelerometer application for the quantification of reflex response. In: 35th Annual International Conference of the IEEE 2013 Jul 3 Engineering in Medicine and Biology Society (EMBC): IEEE; 2013. pp. 4658-4661

[29] LeMoyne R, Mastroianni T, Grundfest W. iPhone wireless accelerometer reflex quantification system using an artificial tendon reflex device. In: Proceedings of 43rd Society for Neuroscience Annual Meeting; San Diego, CA; 2013 Nov

[30] LeMoyne R, Mastroianni T, Montoya K. Implementation of a smartphone for evaluating gait characteristics of a trans-tibial prosthesis. In: 36th Annual International Conference of the IEEE 2014 Aug 26 Engineering in Medicine and Biology Society (EMBC): IEEE; 2014. pp. 3674-3677

[31] LeMoyne R, Mastroianni T. Implementation of a smartphone as a wireless gyroscope application for the quantification of reflex response. In: 36th Annual International Conference of the IEEE 2014 Aug 26 Engineering in Medicine and Biology Society (EMBC): IEEE; 2014. pp. 3654-3657

[32] LeMoyne R, Hessel A, Mastroianni T, Nishikawa K. Implementation of an iPhone as a wireless gyroscope platform for gait analysis. In: Proceedings of 7th World Congress of Biomechanics; Boston, MA; 2014 Jul

[33] LeMoyne R, Mastroianni T, Grundfest W. Wireless accelerometer iPod application for quantifying gait characteristics. In: Annual International Conference of the IEEE 2011 Aug 30 Engineering in Medicine and Biology Society, EMBC: IEEE; 2011. pp. 7904-7907

[34] LeMoyne R, Mastroianni T. iWalk, a gait analysis device incorporating an iPod wireless accelerometer application. In: Proceedings of 42nd Society for Neuroscience Annual Meeting; New Orleans, LA; 2012 Oct 13

[35] LeMoyne R, Mastroianni T. Implementation of an iPod application as a wearable and wireless accelerometer system for identifying quantified disparity of hemiplegic gait. Journal of Medical Imaging and Health Informatics. 2014 Aug 1;**4**(4):634-641

[36] LeMoyne R, Mastroianni T, Coroian C, Grundfest W. Tendon reflex and strategies for quantification, with novel methods incorporating wireless accelerometer reflex quantification devices, a perspective review. Journal of Mechanics in Medicine and Biology. 2011 Jun;**11**(03):471-513

[37] LeMoyne R, Mastroianni T. Reflex response quantification using an iPod wireless accelerometer application. In: Proceedings of 41st Society for Neuroscience Annual Meeting; Washington, DC; 2011 Nov 12

[38] LeMoyne R, Mastroianni T, Grundfest W. Quantified reflex strategy using an iPod as a wireless accelerometer application. In: Annual International Conference of the IEEE 2012 Aug 28 Engineering in Medicine and Biology Society (EMBC): IEEE; 2012. pp. 2476-2479

[39] LeMoyne R, Mastroianni T. Quantification of patellar tendon reflex response using an iPod wireless gyroscope application with experimentation conducted in Lhasa, Tibet and post-processing conducted in Flagstaff, Arizona through wireless Internet connectivity. In: Proceedings of 44th Society for Neuroscience; Washington, DC; 2014 Nov

[40] Hall M, Frank E, Holmes G, Pfahringer B, Reutemann P, Witten IH. The WEKA data mining software: An update. ACM SIGKDD explorations newsletter. 2009 Nov 16;**11**(1):10-18

[41] Witten IH, Frank E, Hall MA. Data Mining: Practical Machine Learning Tools and Techniques; Morgan Kaufmann, Burlington, 2011

[42] WEKA [http://www.cs.waikato.ac.nz/~ml/weka/]

[43] LeMoyne R, Kerr WT, Zanjani K, Mastroianni T. Implementation of an iPod wireless accelerometer application using machine learning to classify disparity of hemiplegic and healthy patellar tendon reflex pair. Journal of Medical Imaging and Health Informatics. 2014 Mar 1;**4**(1):21-28

[44] LeMoyne R, Kerr W, Mastroianni T. Implementation of machine learning with an iPod application mounted to a cane for classifying assistive device usage. Journal of Medical Imaging and Health Informatics. 2015 Dec 1;**5**(7):1404-1408

[45] LeMoyne R, Mastroianni T. Machine learning classification of a hemiplegic and healthy patellar tendon reflex pair through an iPod wireless gyroscope platform. In: Proceedings of 45th Society for Neuroscience; Chicago, IL; 2015 Oct

[46] LeMoyne R, Mastroianni T, Hessel A, Nishikawa K. Ankle rehabilitation system with feedback from a smartphone wireless gyroscope platform and machine learning classification. In: IEEE 14th International Conference on 2015 Dec 9 Machine Learning and Applications (ICMLA): IEEE; 2015. pp. 406-409

[47] LeMoyne R, Tomycz N, Mastroianni T, McCandless C, Cozza M, Peduto D. Implementation of a smartphone wireless accelerometer platform for establishing deep brain stimulation treatment efficacy of essential tremor with machine learning. In: 37th Annual International Conference of the IEEE 2015 Aug 25 Engineering in Medicine and Biology Society (EMBC): IEEE; 2015. pp. 6772-6775

[48] LeMoyne R, Mastroianni T, Tomycz N, Whiting D, McCandless C, Peduto D, Cozza M. I-Phone wireless accelerometer quantification of extremity tremor in essential tremor patient undergoing activated and inactivated deep brain stimulation. Neuromodulation. 2015 Aug 1;**18**(6):e338

[49] LeMoyne R, Mastroianni T. Implementation of a smartphone as a wireless gyroscope platform for quantifying reduced arm swing in hemiplegic gait with machine learning classification by multilayer perceptron neural network. In: IEEE 38th Annual International Conference of the 2016 Aug 16 Engineering in Medicine and Biology Society (EMBC): IEEE; 2016. pp. 2626-2630

[50] Mastroianni T, LeMoyne R. Application of a multilayer perceptron neural network with an iPod as a wireless gyroscope platform to classify reduced arm swing gait for people with Erb's palsy. In: Proceedings of 46th Society for Neuroscience; San Diego, CA; 2016 Nov

[51] LeMoyne R, Mastroianni T. Smartphone wireless gyroscope platform for machine learning classification of hemiplegic patellar tendon reflex pair disparity through a multilayer perceptron neural network. In: Wireless Health (WH): IEEE; 2016 Oct. pp. 103-108

[52] LeMoyne R, Mastroianni T. Implementation of a multilayer perceptron neural network for classifying a hemiplegic and healthy reflex pair using an iPod wireless gyroscope platform. In: Proceedings of 46th Society for Neuroscience; San Diego, CA; 2016 Nov

[53] LeMoyne R, Mastroianni T. Wireless gyroscope platform enabled by a portable media device for quantifying wobble board therapy. In: IEEE 39th Annual International Conference of the 2017 (pending publication) Engineering in Medicine and Biology Society (EMBC): IEEE; 2017

[54] LeMoyne R, Mastroianni T. Virtual Proprioception for eccentric training. In: IEEE 39th Annual International Conference of the 2017 (pending publication) Engineering in Medicine and Biology Society (EMBC): IEEE; 2017

[55] LeMoyne R, Coroian C, Mastroianni T, Grundfest W. Virtual proprioception. Journal of Mechanics in Medicine and Biology. 2008 Sep;**8**(03):317-338

[56] LeMoyne R, Coroian C, Mastroianni T, Wu W, Grundfest W, Kaiser W. Virtual proprioception with real-time step detection and processing. In: 30th Annual International Conference of the IEEE Engineering in Medicine and Biology Society 2008 (EMBS 2008): IEEE; 2008 Aug 20. pp. 4238-4241

[57] Rebula JR, Ojeda LV, Adamczyk PG, Kuo AD. Measurement of foot placement and its variability with inertial sensors. Gait & Posture. 2013 Sep 30;**38**(4):974-980

[58] Madgwick S. An efficient orientation filter for inertial and inertial/magnetic sensor arrays. Report x-io. UK: University of Bristol; 2010 Apr 30. pp. 1-32

[59] Madgwick SO, Harrison AJ, Vaidyanathan R. Estimation of IMU and MARG orientation using a gradient descent algorithm. In: IEEE International Conference on 2011 Jun 29 Rehabilitation Robotics (ICORR): IEEE; 2011. pp. 1-7

[60] Cognition Engineering [http://cognitionengineering.com]

Chapter 2

The Geometry and Usage of the Supplementary Fisheye Lenses in Smartphones

Cumhur Sahin

Additional information is available at the end of the chapter

http://dx.doi.org/10.5772/intechopen.69691

Abstract

Nowadays, mobile phones are more than a device that can only satisfy the communication need between people. Since fisheye lenses integrated with mobile phones are lightweight and easy to use, they are advantageous. In addition to this advantage, it is experimented whether fisheye lens and mobile phone combination can be used in a photogrammetric way, and if so, what will be the result. Fisheye lens equipment used with mobile phones was tested in this study. For this, standard calibration of 'Olloclip 3 in one' fisheye lens used with iPhone 4S mobile phone and 'Nikon FC-E9' fisheye lens used with Nikon Coolpix8700 are compared based on equidistant model. This experimental study shows that Olloclip 3 in one fisheye lens developed for mobile phones has at least the similar characteristics with classic fisheye lenses. The dimensions of fisheye lenses used with smart phones are getting smaller and the prices are reducing. Moreover, as verified in this study, the accuracy of fisheye lenses used in smartphones is better than conventional fisheye lenses. The use of smartphones with fisheye lenses will give the possibility of practical applications to ordinary users in the near future.

Keywords: smartphone, fisheye lenses, equidistant projection model, distortion geometry, supplementary lenses

1. Introduction

Nowadays, mobile phones are more than a device that can only satisfy the communication need between people. Cameras and other integrated additional devices are found in almost every smartphone. Other than these devices, there are tele, macro and fisheye lenses that can easily be integrated to the smartphones. Some of those lens kits are presented in Refs. [1, 2]. Fisheye lenses that are compliant to mobile phones are one of these new equipments. Since fisheye lenses integrated with mobile phones are lightweight and easy to use, they are advantageous. Additionally, these lenses are cost efficient compared to conventional fisheye lenses. The characteristics of Olloclip lens used in this study are presented in Ref. [3]. Cameras on mobile phones are as capable as compact cameras that we use in our daily lives. Smartphone cameras used for acquiring image instead of conventional cameras have opened a new scientific study field. Another scientific study field is that using smartphone cameras together with the developing technologies has given the opportunity to achieve new study fields that have not been made before. Chugh et al. [4] present a detailed survey of methods for detecting road conditions. Smartphone sensors are gaining importance in this field, as they are cost effective and also increase scalability. Analysing from the research activities, it is certain that this area will gain more importance in recent future. The objective of the research is to improve traffic safety through collecting and distributing up-to-date road surface condition information using mobile phones [5]. Perttunen et al. [5] present experimental results from real urban driving data that demonstrate the usefulness of the system. To monitor road and traffic conditions in such a setting, Mohan et al. [6] present Nericell, a system that performs rich sensing by piggybacking on smartphones that users carry with them in normal course. Mohan et al. [6] focus specifically on the sensing component, which uses the accelerometer, microphone, GSM radio and/or GPS sensors in these phones to detect potholes, bumps, braking and honking. Wagner et al. [7] present two techniques for natural feature tracking in real-time on mobile phones and use an approach based on heavily modified state-of-the-art feature descriptors, namely scale invariant feature transform (SIFT) and Ferns. Object-wise 3D reconstruction is a cardinal problem in computer vision, with much work being dedicated to it throughout recent years. Unlike other approaches, some approaches use global computation, whereas Prisacariu et al. [8] adopt a local computation method related with signed distance transformation and its derivatives. By this method, 3D renderings are quickly obtained by hierarchical ray casting. Real-time mobile phone performances and speeds faster than 100 fps on PC are achieved by the tracker and GPU acceleration is not required [8]. Tanskanen et al. [9] propose the first dense stereo-based system for live interactive 3D reconstruction on mobile phones. Pan et al. [10] present a novel system that allows for the generation of a coarse 3D model of the environment within several seconds on mobile smartphones. The contribution of this work is the presentation of a novel approach to generate visually appealing, textured 3D models from a set of at least three panoramic images on mobile phones without the need for remote processing [10]. Wagner et al. [11] present a novel method for the real-time creation and tracking of panoramic maps on mobile phones. The maps generated with this technique are visually appealing, very accurate and allow drift-free rotation tracking. Nowadays, smartphones are widely used in the world, and generally, they are equipped with many sensors. Almazan et al. [12] study how powerful the low-cost embedded Inertial Measurement Unity (IMU) and Global Positioning System (GPS) could become for intelligent vehicles. Main contribution is the method employed to estimate the yaw angle of the smartphone relative to the vehicle co-ordinate system. The results show that the system achieves high accuracy, the typical error is 1%, and is immune to electromagnetic interference [12]. Recently, mobile phones have become increasingly attractive for augmented reality (AR). The recent advent of GPS and orientation sensors on commodity mobile devices has led to the development of numerous mobile augmented reality (AR) applications and broader public awareness and use of these applications. By using the phone orientation sensor to display the appropriate subset of the panorama, orientation accuracy can be effectively increased and augmentations tightly registered with the background [13]. Kurz and Benhimane [14] presented novel approaches to use the direction of the gravity measured with inertial sensors to improve different parts in the pipeline of handheld AR applications [14]. Amongst all the possible applications, AR systems can be very useful as visualization tools for structural and environmental monitoring. Porzi et al. [15] presented a successful implementation on an android device of an egomotion estimation algorithm by porting the tracking module of parallel tracking and mapping (PTAM). Porzi et al [15] described the development of the egomotion estimation algorithm for an android smartphone. In recent decades, many indoor positioning techniques have been researched and some approaches have even been developed into consumer products. Two devices are selected, the iPhone 3GS and the iPhone 4, to analyse their sensors for usability of an inertial navigation system. A precise Inertial Navigation System (INS) cannot be completely acquired by a strapdown algorithm because of inaccurate and noisy sensors that are used by both the iPhones. In order to enhance the accuracy, several filters were used. Finally, strapdown algorithms were analysed and verified with related testing and best filter combination was found for each of the devices [16]. Burgess et al. [17] expand on previous work by using a multi-floor model taking into account dampening between floors, and optimize a target function consisting of least squares residuals, to find positions for WiFis and the smartphone measurement locations [17]. Burgess et al. [18] have presented a method for simultaneously mapping the radio environment and positioning several smartphones in multi-story buildings.

Computer vision applications for mobile phones are gaining increasing attention due to several practical needs resulting from the popularity of digital cameras in today's mobile phones. Hadid et al. [19] described the task of face detection and authentication in mobile phones, and experimentally analyse a face authentication scheme using Haar-like features with AdaBoost for face and eye detection and local binary pattern (LBP) approach for face authentication. Shen et al. [20] address the challenges of performing face recognition accurately and efficiently on smartphones by designing a new face recognition algorithm called opti-sparse representation classification (opti-SRC). Sparse representation classification (SRC) is a state-of-the-art face recognition algorithm, which has been shown to outperform many classical face recognition algorithms in OpenCV.

Monitoring aquatic environment is of great interest to the ecosystem, marine life and human health [21]. An efficient method for monitoring marine debris is smartphone-based aquatic robot (SOAR). It is a robotic system having low cost. The aim is to monitor debris in water environment. It contains a smartphone and a robotic fish platform. Robotic fish have a capability to moving through water and smartphone is used to capture images [22]. Another method for detecting debris is Samba. Samba is an aquatic robot that contains a smartphone and a robotic fish platform to monitor harmful marine debris. Using camera of the smartphone, Samba can recognize aquatic debris in dynamic and complex environments [22]. Maindalkar and Ansari [23] present design of aquatic robot for aquatic pollutants monitoring. The android smartphone is integrated with aquatic robot to capture images and to acquire data of different sensors. The implemented design contains CV algorithm for image processing on openCV platform. The real-time pollutant detection is done with the CV algorithm efficiently [23].

Muaremi [24] investigate the potential of a modern smartphone and a wearable heart rate monitor for assessing affect changes in daily life. Muaremi et al. [24] use smartphone features and heart rate variability (HRV) measures as predictors for building classification models to discriminate among low, moderate and high perceived stress. As smartphones evolve, researchers are studying new techniques to ease the human-mobile interaction. User interface of mobile phone can be operated by eye tracking and blink detection functions on EyePhone. These results are preliminary, but they suggest that EyePhone is a favourable tool for driving mobile applications with automation [25]. The advent of mobile sensing technology provides a potential solution to the challenge of collecting repeated information about both behaviours and situations such as to detect the type of situation using the sensors built into today's ubiquitous smartphones [26]. Sandstrom et al. [26] focused on using location sensors to learn the semantics of places, so that we could examine relationships between place, affect and personality. Sensor-enabled smartphones are opening a new frontier in the development of mobile sensing applications. The recognition of human activities and context from sensor-data using classification models underpins these emerging applications [27]. The key contribution of community similarity networks (CSN) is that it makes the personalization of classification models practical by significantly lowering the burden to the user through a combination of crowd-sourced data and leveraging networks that measure the similarity between users. Lu et al. [28] present Jigsaw, a continuous sensing engine for mobile phone applications that require continuous monitoring of human activities and context. Supporting continuous sensing applications on mobile phones is very challenging. Lu et al. [29] propose StressSense for unobtrusively recognizing stress from human voice using smartphones. Lane et al. [30] discuss the emerging sensing paradigms, and formulate an architectural framework for discussing a number of open issues and challenges emerging in the new area of mobile phone sensing research [30]. Rachuri et al. [31] have presented EmotionSense, a novel system for social psychology study of user emotion based on mobile phones. Rachuri et al. [31] have presented the design of novel components for emotion and speaker recognition based on Gaussian mixture models. The driving vision is a smartphone service, called MoodSense, that can infer its owner's mood based on information already available in today's smartphones. In Ref. [32], it is suggested that user mood can be separated into four main types with 91% average accuracy. These results can be obtained with 3 weeks of research data and basic smartphone handling statistics. Although these results are not decisive, they show practicability of mood inference without any microphone and/or camera with bulky power requirements and social interaction [32].

Recently, the calibration methods using display devices such as monitors, tablets or smartphones have come to the forefront [33]. Gruen and Akca [34] report about first experiences in calibration and accuracy validation of mobile phone cameras. Ha et al. [33] propose a novel camera calibration method for defocused images using a smartphone under the assumption that the defocus blur is modelled as a convolution of a sharp image with a Gaussian point spread function (PSF). The effectiveness of the proposed method has been emphasized in several real experiments using a compact display device such as a smartphone [33]. Delaunoy et al. [35] propose a new approach to estimate the geometric extrinsic calibration of all the elements of a smartphone or tablet (such as the screen, the front and the back cameras) by using a planar mirror. Saponaro and Kambhamettu [36] described a method for calibrating a smartphone camera by taking two images at different rotations while tolerating small translations. Ahn et al. [37] were intended to analyse accuracy of smartphone image in determining three-dimensional location for approximated objects before photo survey system using smartphone is developed, and then evaluate its usability.

Fisheye lenses provide instant wide-angle images from one point with a single camera. Fisheye optics are placed onto charge couple device (CCD) or complementary metal oxide semiconductor (CMOS) cameras without requiring any complex technology. They do not require an external mirror or rotational device. Thus, these optics are small in size and do not require any maintenance [38]. They have a very short focal length, which produces a hemisphere [39]. By using fisheye lenses, a large area of any surrounding space can be acquired with a single photograph. Therefore, fisheye lenses are useful in most of the applications. In addition to high quality landscape and interior visualizations (e.g. ceiling frescos of historical buildings) in commercial demonstrations or internet presentations, fisheye images are also beneficial for measurement operations [40].

The first fisheye lenses have been created by Hill in 1924 [41], but, they have not been preferred in photogrammetric measurements since they provide images with huge distortions and they do not meet central projection. Using the images obtained from fisheye lens imaging systems in photogrammetric measurement and modeling processes becomes popular in recent years by the help of the development in software and hardware technologies. Later, a significant increase has been seen in terms of volume scientific research on this subject matter. Recently, there have been several academic studies presenting the benefit from fisheye lenses. Fisheye cameras are finding increasing number of applications in surveillance, robotic vision, automotive rear-view imaging systems, etc. because of their wide-angle properties [42]. Fisheye lens cameras have also been used during sky observations [43], visual sun compass creation [44], and sunpath diagram derivation [45]. Beekmans et al. [46] present a complete approach for stereo cloud photogrammetry using hemispheric sky imagers. This approach combines calibration, epipolar rectification and block-based correspondence search for dense fisheye stereo reconstruction for clouds. A novel panoramic imaging system that uses a curved mirror as a simple optical attachment to a fisheye lens is given in Ref. [47]. Streckel et al. [48] describe a visual markerless real-time tracking system for augmented reality applications. The system uses a firewire camera with a fisheye lens mounted at 10 fps. Brun et al. [49] present a new mobile mapping system mounted on a vehicle to reconstruct outdoor environment in real time. Yamamoto et al. [50] propose a mobile web map interface that is based on a metaphor of the wired fisheye lens. The user can easily navigate through the area surrounding the present location while keeping the focus within the map. These features enable users to find the target quickly. Yamamoto et al. [50] confirmed the advantages of the proposed system by evaluation experiments. The new system will be able to contribute to the novel mobile web map services with fisheye views for mobile terminals such as cellular phones. Ahmad and Lima [51] present a cooperative approach for tracking a moving spherical object in three-dimensional space by a team of mobile robots equipped with sensors in a highly dynamic environment. Zheng and Li [52] explore the use of a fisheye camera to achieve the scene tunnel acquisition. In Ref. [53], authors have focused on dioptric systems to implement a robot surveillance application for fast and robust tracking of moving objects in dynamic, unknown environments. Another application that uses fisheye lens is a research that examines the use of fisheye lenses as optical sensors on unmanned aerial vehicle (UAV) platform in Queensland Technical University in Australia [54]. Grelsson [55] used a fisheye camera for horizon detection in aerial images. Naruse et al. [56] propose three-dimensional measurement method of underwater objects using a fisheye stereo camera. In Ref. [57], a novel technique to accurately estimate the global position of a moving car using an omnidirectional camera and untextured three-dimensional city model is proposed. Today, one of the areas that most frequently benefit from fisheye lenses is applications done in combination with terrestrial laser scanners. Georgantas et al. [58] present a comparison of automatic photogrammetric techniques to terrestrial laser scanning for three-dimensional modeling of complex interior spaces. The 8 mm fisheye lens that was used allowed us to acquire photos with a global view of the scene and thus with textured zones in every image, which is essential for the scale invariant feature transform (SIFT) algorithm. Image analysis tasks such as 3D reconstruction from endoscopic images require compensation of geometric distortions introduced by the lens system [59]. Hu et al. [60] propose effective pre-processing techniques to ensure the applicability of face detection tools onto highly distorted fisheye images.

Schneider and Schwalbe [61] present the integration of a geometric model of fisheye lenses and a geometric terrestrial laser scanner model in a bundle block adjustment. Fisheye projection functions are designed such that a greater portion of the scene is projected onto the image sensor on the image plane, at the expense of introducing (often considerable) radial distortion [62]. The fisheye lens camera should be calibrated to be used in applications that require high accuracy [63]. There are different studies in literature, which focus on the calibration of fisheye lenses. Abraham and Forstner [38] presented rigorous mathematical models for the calibration of a stereo system composed of two fisheye lens cameras and for the epipolar rectification of the images acquired by this dual system.

Arfaoui and Thibault [64] have described a method using a compact calibration object for fisheye lens calibration. The setup generated a robust and accurate virtual calibration grid, and the calibration was performed by rotating the camera around two axes. The experimental results and the comparison with a 3D calibration object showed that the virtual grid method is efficient and reliable [64]. Kim and Paik [65] presented a novel 3D simulation method for fisheye lens distortion in a vehicle rear-view camera. The proposed method creates a geometrically distorted image of an object in 3D space according to the lens specifications. The proposed simulation method can be applied to designing a general optical imaging system for intelligent surveillance as well as a vehicle rear-view backup camera [65] Torii et al. [66] present a pipeline for camera pose and trajectory estimation, and image stabilization and rectification for dense as well as wide baseline omnidirectional images. The experiments with real data demonstrate the use of the proposed image stabilization method. Five image sequences of a city scene captured by a single hand-held fisheye lens camera are used as our input [66].

In Ref. [67], Kodak DSC 14 Pro with Nikkor 8 mm fisheye lens is calibrated with an equidistant projection. In addition to decentring, symmetric radial and affinity distortion models, precise mathematical models were used, which were based on stereo-graphic, equidistant, orthogonal and equisolid-angle projections. Kannala and Brandt [68] propose a generic camera model, which is suitable for fisheye lens cameras as well as for conventional and wide-angle lens cameras, and a calibration method for estimating the parameters of the model. Fisheye lenses are not perspective lenses, image resolution in these lenses are not fixed (univocal), illumination is not distributed homogeneously [69]. Upto now, many researchers have considered the relationship between distorted radius and undistorted radius in the image plane ignoring the variation of the angle. Zhu et al. [70] present a fisheye camera model based on the refractive nature of the incoming rays and estimate the model parameters without calibration objects using Micusik's method [71]. In photogrammetry, the collinearity mathematical model, based on perspective projection combined with lens distortion models, is generally used in the camera calibration process. However, fisheye lenses are designed for the following different spherical projections models such as stereographic, equidistant, orthogonal and equisolid angle [63]. The calibration results of Fuji-Finepix S3pro camera with Bower-Samyang 8 mm lens were assessed by the help of precise mathematical models. Bower-Samyang 8 mm is cheaper than other fisheye lenses and unlike others; it is based on stereographic projection [63].

Most of the fisheye lenses are technically based on equidistant or equisolid-angle projection. Initially, equisolid-angle projection geometry is constructed and then diagonal fisheye lenses are constructed. The distortion of the image edges is more significant than fisheye lenses with equidistant projection. The only way to construct orthographic projection geometry is to use sophisticated optical construction. Stereographic projection is not practically realizable [67]. Among the other models proposed, an important one is the equidistant model. The model proposes that the distance between an image point and the centre of radial distortion is proportional to the angle between a corresponding three-dimensional point, the optical centre and the optical axis [72]. Equidistant fisheye lenses are often used for scientific measurement where the measurement of angles is necessary. Thus, it is also sometimes referred to as an equiangular fisheye lens [73]. Perhaps the most common model is the equidistance projection [68]. Friel et al. [74] use the equidistance projection equation to describe the radial distortion, as this is typically among the most commonly used and inexpensive fisheye lens types. The work described in Ref. [74] shows that it is possible to carry out automatic calibration of fisheye lenses, using information derived from real-world automotive scenes, and to obtain calibration data to a high degree of accuracy.

The main purpose of this study is to test fisheye lens equipment used with mobile phones. Mobile phone imaging with the additional hardware has been used more popularly not only outside but also in indoor applications. Therefore, hardware properties of this wide-angle optics will be used in the photogrammetric documentation in the near future for mobile phone imaging. Since fisheye lenses integrated with mobile phones are lightweight and easy to use, they are advantageous. In addition to this advantage, it is experimented whether fisheye lens and mobile phone combination can be used in a photogrammetric way, and if so, what will be the result. In this study, standard calibration of 'Olloclip 3 in one' fisheye lens used with iPhone 4S mobile phone and 'Nikon FC-E9' fisheye lens used with Nikon Coolpix8700 are compared based on equidistant model. By using photogrammetric bundle block adjustment, the results of these calibrations are analysed. Geometric properties of these wide-angle lenses will be more important in the photogrammetric measurement assessment. This study suggests a pre-calibration process of these kinds of hardware for the photogrammetric process in the test field. In the literature, although there are many geometric camera calibration publications, none of them compares the mobile phone fisheye lens kit with conventional fisheye lens on the fundamentals of photogrammetric measurement assessment. The results of this photogrammetric process are also compared with conventional wide-angle hardware in this paper.

The second section of this chapter briefly describes fisheye projection models. The third section of this chapter briefly describes equidistant model. The fourth section reports an empirical study for calibration of the combination of iPhone 4S camera with Olloclip 3 in one fisheye lens and Nikon Coolpix8700 camera FC-09 fisheye lens combination by using equidistant model. The fifth section interprets the results that resulted from the experiment process. The sixth section concludes the study.

2. Fisheye projection models

Pinhole projection is so called because it preserves the rectilinearity of the projected scene (i.e. straight lines in the scene are projected as straight lines on the image plane). The Pinhole (perspektife) projection is shown in **Figure 1**. The Pinhole (perspektife) projection mapping function is given in Eq. (1).

$$r_u = f.\tan(\theta)(\text{perspective projection}) \tag{1}$$

where f is the distance between the principal point and the image plane, θ is the incident angle (in radians) of the projected ray to the optical axis of the camera and r_u is the projected radial distance from the principal point on the image plane. However, for wide field of view (FOV) cameras, under rectilinear projection, the size of the projected image becomes very large, increasing to infinity at an FOV of 180° [62].

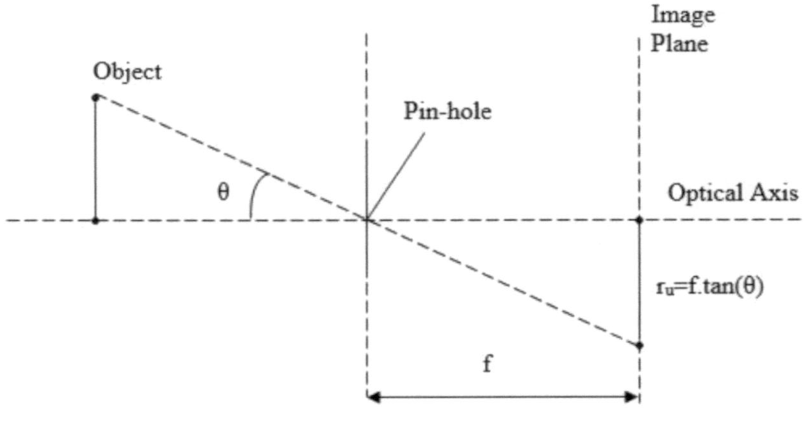

Figure 1. Pinhole (perspektife) projection representation.

Interior orientation parameters (IOPs) can be estimated by a procedure called camera calibration. The perspective bundle, which generated the image, can be reconstructed by this procedure. The principal point co-ordinates, focal length and coefficient for systematic errors correction (lens distortion: symmetric radial and decentring and affinity) are the IOPs of digital cameras. When additional parameters (IOPs) in Eq. (2) [75] are examined, collinearity equations are the most popular camera calibration method [63, 76].

$$\begin{aligned} x_f &= x' - x_0 - \Delta x = -f\frac{X_c}{Z_c} \\ y_f &= y' - y_0 - \Delta y = -f\frac{Y_c}{Z_c} \end{aligned} \tag{2}$$

where, f represents the focal length, and (X_c, Y_c, Z_c) shows the 3D point co-ordinates of photogrammetric reference system in Eq. (3); point co-ordinates of the image are (x_f, y_f); image point co-ordinates of the reference system parallel to photogrammetric system are represented as (x', y'), this element originates from image centre and principal point (pp) of the co-ordinates are (xo, yo).

$$\begin{aligned} X_c &= r_{11}.(X - X_{CP}) + r_{12}.(Y - Y_{CP}) + r_{13}.(Z - Z_{CP}) \\ Y_c &= r_{21}.(X - X_{CP}) + r_{22}.(Y - Y_{CP}) + r_{23}.(Z - Z_{CP}) \\ Z_c &= r_{31}.(X - X_{CP}) + r_{32}.(Y - Y_{CP}) + r_{33}.(Z - Z_{CP}) \end{aligned} \tag{3}$$

where r_{ij} (i and j from 1 to 3) represents rotation matrix elements and with r_{ij}, the object can be used in relation to the image reference system; (X, Y, Z) shows any point's co-ordinates in the object reference system and (X_{cp}, Y_{cp}, Z_{cp}) shows perspective centre (PC) in object reference system [63]. Pinhole (perspektife) projection model is not suitable for fisheye lenses. Fisheye lenses instead are usually designed to obey one of the following projections [68]:

$$r = 2.f.\tan(\theta/2) \text{ (stereographic projection)} \tag{4}$$

$$r = f.\theta \text{ (equidistant projection)} \tag{5}$$

$$r = 2.f.\sin(\theta/2) \text{ (equisolid angle projection)} \tag{6}$$

$$r = f.\sin(\theta) \text{ (orthogonal projection)} \tag{7}$$

In Eqs. (1) and (4)–(7), the angle between optical axis and incoming ray is shown with θ symbol; the distance between image point and principal point is represented with r, and focal length is represented with f. Equidistance projection can be accepted as the most wide-spread used fisheye lens model. **Figure 2a** illustrates the schematic description of different projections for the fisheye lens. **Figure 2b** shows the difference between pinhole lens and fisheye lens. The images acquired with non-perspective projection are more near to principal point when the results are compared to the results of perspective projection. Therefore, the view angle of fisheye lens is wider than conventional lens. Moreover, actual image surface of fisheye lens presents a hemisphere in accordance with a pinhole lens plane. Thus, projecting the image on surface of the hemisphere into an actual imaging plane results in a deformation of the fisheye lens [77].

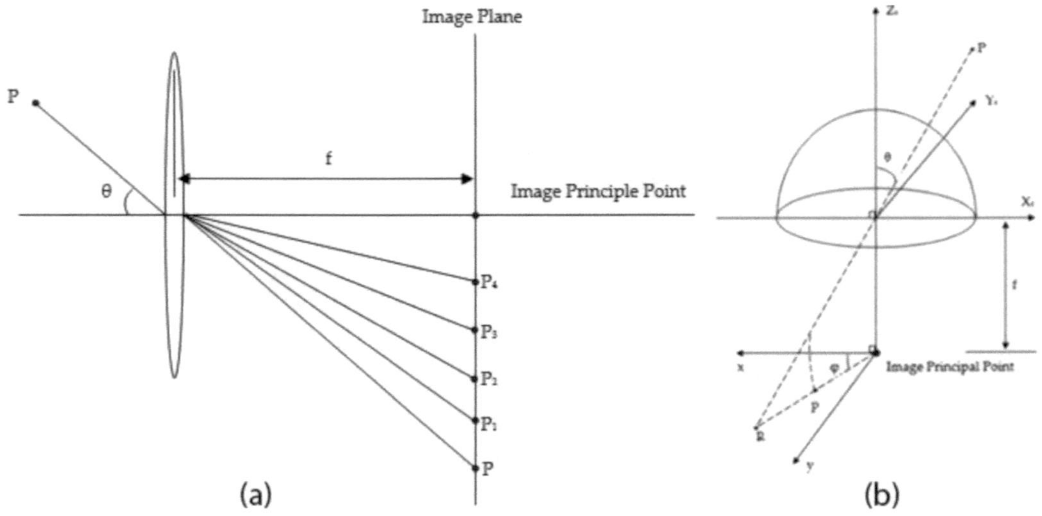

Figure 2. The principles for various lenses: (a) shows different lens projections, p, p_1, p_2, p_3 and p_4 are respectively perspective projection, stereographic projection, equidistance projection, equisolid angle projection and orthogonal projection; the corresponding distances between image points and the principal point are represented with r, r_1, r_2, r_3 and r_4; (b) shows the difference between pinhole lens and fisheye lens. In terms of fisheye lens, perspective image's projection on the hemisphere surface into the image plane is the actual image.

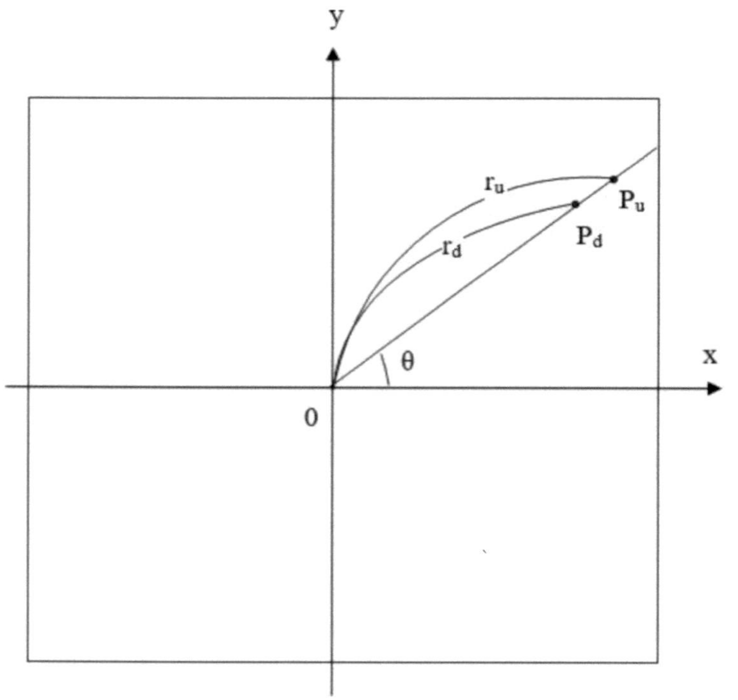

Figure 3. Radial distortion in the 2D imaging plane: O represents the image centre, P_u, $P_d \in R^2$ are pixel co-ordinate vectors in the input undistorted and output distorted images, respectively, r_u, $r_d \in$ R are the distance of P_u and P_d from centre, and $\theta = \theta_u = \theta_d$is the angle of OP_u or, equivalently, OP_d.

http://dx.doi.org/10.5772/intechopen.69691

A wide-angle lens produces geometric distortion in the radial direction called the barrel distortion, since it compresses the peripheral region to contain a wide angle of view in the image plane. Considering this problem, many researchers have proposed various models to correct the barrel distortion of the wide-angle lens. A two-dimensional (2D) approximated barrel distortion model is shown in **Figure 3**, where an original pixel P_u moves towards the centre at P_d along the radial direction in the image plane [65]. A polynomial model was proposed to approximate various types of wide-angle lenses using the distortion coefficients. The distance of the distorted pixel P_u is determined by the polynomial equation [65].

3. Equidistant projection function for fisheye lenses

In order to model the perfect fisheye lens, scene projections are necessary. These can be defined by two main characteristics. Firstly, field of vision covers 2π steradians, it creates a circular image and the distortions become symmetrical with reference to centre of the image. Secondly, fisheye lens has an infinite depth of field. All objects in the image have a precise focus. Therefore, two postulates, namely the azimuth angle invariability and the equidistant projection rule, govern the formation of non-linear image distortion. These pre-suppositions explain the projection of object points into the sensor. They directly affect the eventually developing dewarping algorithm [78].

The azimuth angle invariability, which is the first postulate, determines the projection of points of the plane (which passes through the optical axis that is perpendicular to the sensor plane). The azimuth angle of the object points and their projections onto the sensor remain unchanged due to differences in the object distance or elevation within the content plane [78]. According to Ref. [79], the equidistant lens is 'preferable for measurement of incidence angles (θ) and azimuth angles. The effect of error of lens position is small, and the linear relation of radial distance (r_d) and incidence angle (θ) of a ray from the three-dimensional point is convenient to analyse'.

The second postulate, the equidistant projection rule, depicts the relationship between radial distance (r_d) of an image point on the sensor plane-zenith (incidence (θ)) angle which is created by the vector of image centre-world object point in **Figure 4**. According to this rule, there is a linear relationship between the centre to r_d image point radial distance and (θ) zenith angle [78].

As the zenith angle varies from 0 to 90°, the radial distance of the corresponding image point varies linearly from 0 to a maximum value R, determined by the modelled sphere's [78].

(r_d) on the image plane, which is the radial distance in equidistant projection, is directly proportional to incident ray's angle. It is equivalent to arc segment's length, which is located between the z-axis and the projection ray of point P on the sphere in **Figure 5** [62].

Thus, the equidistant projection function is given in Eq. (8).

$$r_d = f.\theta \tag{8}$$

where rd is the fisheye radial distance of a projected point from the centre, f is focal distance and θ represents the incidence angle of a ray which begins from the projected three-dimensional

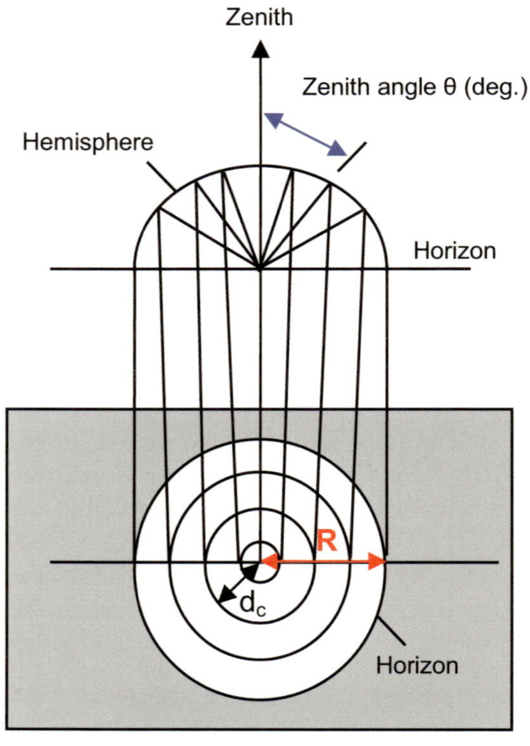

Figure 4. Equidistant projection (Equidistant projection, $\theta/90 = d_c/R$) [43].

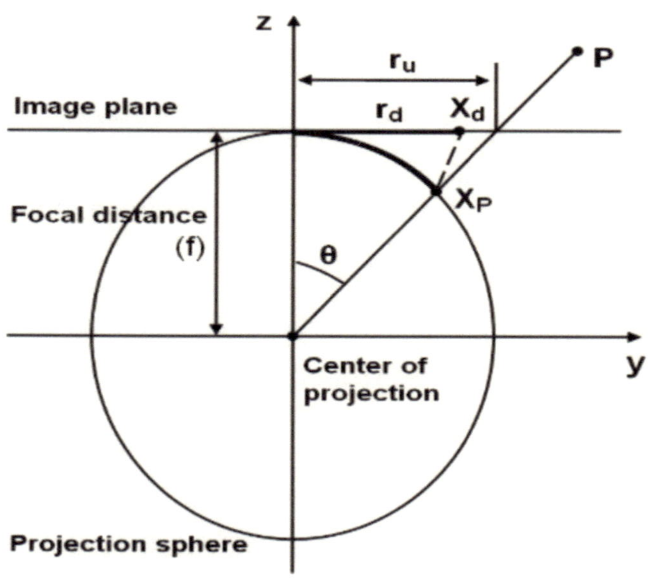

Figure 5. Equidistant fisheye projection function representation.

http://dx.doi.org/10.5772/intechopen.69691

point into the image plane. In fisheye cameras, following process is performed by the help of this common mapping function. The other mapping functions are stereographic, equisolid and orthogonal [80]. Eq. (9) is derived by substituting arctangent function for θ in Eq. (8). Where r_u is the height of the projection on the image plane (the subscript u being used to denote the undistorted projection) [73].

$$r_d = f.\arctan\left(\frac{r_u}{f}\right) \tag{9}$$

In equidistant projection model, the distorted radial distance on the image plane is linearly expressed as the projected ray's angle in radians. Moreover, the length of the arc segment between z-axis and x_p is equivalent to the projected distorted distance r_d (x_p is the intersection point of the projection ray of point X, which has the projection sphere) [73].

Most real optical systems have some undesirable effects, rendering the assumption of the pinhole camera model inaccurate. The most evident of these effects is radial barrel distortion, particularly noticeable in fisheye camera systems, where the level of this distortion is relatively extreme [62]. For most of the applications, the effect of radial distortion can be negligible in normal and narrow field of view (FOV) cameras. However, radial distortion can cause some problems in wide-angle and fisheye cameras both in terms of visual issues and in the processing of computer vision applications such as object detection, recognition and classification processes [73]. Because of the distortion of the radial lens, points on the image plane are displaced from their ideal position into rectilinear pinhole camera model in a non-linear way. The movement occurs in a radial axis from distortion centre on the equidistant image plane. The image in the foveal areas has a better resolution because of the displacement factor of fisheye optics. In addition, the peripheral areas of the image satisfy a resolution that decreases non-linearly [81].

Additional parameters to compensate for deviations of the geometric fisheye model from the physical reality are the same parameters that are applied, as they are in common use, for central perspective lenses [69]. Accordingly, the equidistant projection function with additional parameters is given in Eq. (10).

$$r_d = f.\arctan\left(\frac{r_u}{f}\right) + \text{Additional Parameters} \tag{10}$$

Due to the particularly high levels of distortion present in fisheye cameras, there have been several alternative models developed [81]. Some models are fisheye transform, field of view, division model, and polynomial model [82]. The work in Ref. [67], investigates the addition of the brown-parameters to the basic geometric fisheye model to compensate for the remaining 'systematic effects' [82].

Three co-ordinate systems are used in order to define the projection of an object point into a hemispherical fisheye-dimensional image. These are: the superordinated cartesian object co-ordinate system (X, Y, Z) and the camera co-ordinate system (x, y, z) in **Figure 6**. The image co-ordinate system (x', y') is defined similar to its usual definition in photogrammetric applications. So, the image centre becomes the origin. The x' and y' axes are parallel with the x and y axes of

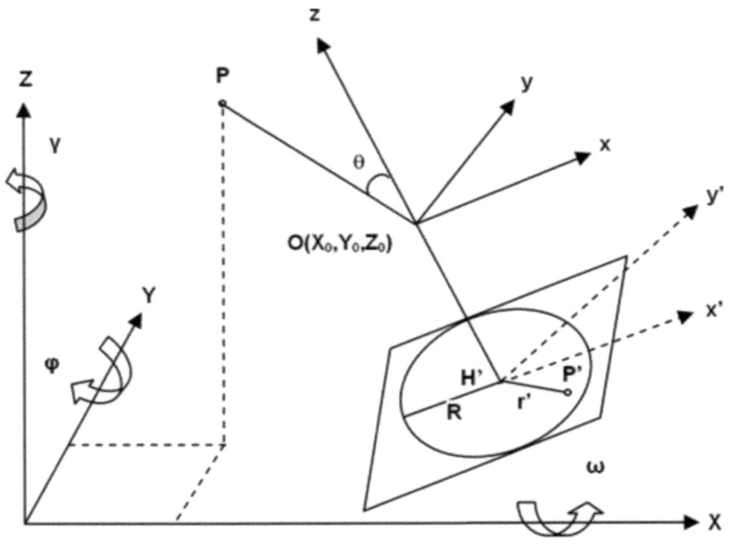

Figure 6. Geometrical model of a fisheye camera.

camera co-ordinate system [67]. The geometric concept is based on the dependence of the image radius r' and the angle of incidence θ [61].

Object co-ordinates are transformed into the camera co-ordinate system. Eq. (11), where X is the co-ordinate vector in the object co-ordinate system, x is the co-ordinate vector in the camera co-ordinate system, R is the rotation matrix and X_0 is the translation between object and camera co-ordinate system:

$$x = R^{-1}(X - X_0) \tag{11}$$

The incidence angle θ in the camera co-ordinate system is defined as follows:

$$\tan\theta = \frac{\sqrt{x^2 + y^2}}{z} \tag{12}$$

Instead of functions for the image radius r', functions for the image co-ordinates x' and y' are required. For this purpose, Eq. (13) is applied:

$$r' = \sqrt{x'^2 + y'^2} \tag{13}$$

After transformations of the equations described above, the final fisheye projection equations for the image co-ordinates is derived. The model equations are finally extended by the co-ordinates of the principal point x'_0 and y'_0 Eq. (14) and the correction terms $\Delta x'$ and $\Delta y'$ [Eqs. (15) and (16)], which contain additional parameters to compensate for systematic effects.

http://dx.doi.org/10.5772/intechopen.69691

Equidistant projection:

$$x' = c.\frac{\arctan\frac{\sqrt{x^2+y^2}}{z}}{\sqrt{\left(\frac{y}{x}\right)^2+1}} + x'_0 + \Delta x' \qquad y' = c.\frac{\arctan\frac{\sqrt{x^2+y^2}}{z}}{\sqrt{\left(\frac{x}{y}\right)^2+1}} + y'_0 + \Delta y' \tag{14}$$

$$\Delta x = x'.\left(A_1 r'^2 + A_2 r'^4 + A_3 r'^6\right) + B_1.\left(r'^2 + 2x'^2\right) + 2B_2 x'y' + C_1.x' + C_2.y' \tag{15}$$

$$\Delta_y = y'.\left(A_1 r'^2 + A_2 r'^4 + A_3 r'^6\right) + 2B_1 x'y' + B_2.(r'^2 + 2y'^2) \tag{16}$$

where;

A_1, A_2, and A_3 are radial distortion parameters,

B_1 and B_2 are decentric distortion parameters,

C_1 and C_2 are horizontal scale factor and shear factor, respectively, and

c is the camera constant, which equals to focal distance.

4. Experiments

The characteristics of the cameras and fisheye lenses, which were chosen for the application, are given below.

Nikon Coolpix8700 digital camera has 8 megapixels resolution and CCD sensor. The 8x optical Zoom-Nikkor lens (f/2.8 – 4.2) offers a focal range of 8.9–71.2 mm [83]. Nikon FC-E9 fisheye lens: focal length of the camera's lens reduced to x0.2. Provides approximate 183° (COOLPIX 5700)/190° (COOLPIX 5400) view angle [83].

iPhone 4S camera has 8 megapixels resolution and CMOS sensor. Its focal length is 35 mm [84]. Olloclip 3 in one fisheye lens: The Olloclip is a device providing three different lens options for iPhone, these are wide-angle, fisheye and macro. The Olloclip with the fisheye lens acquires 180° field of view [85]. **Table 1** shows technical specifications of the cameras used in the application.

The Olloclip 3 in one was mounted on the iPhone 4S. Images acquired with iPhone 4S and Olloclip 3 in one fisheye lens combination were captured with a focal distance of 4.28 mm. Images have 3264 × 2448 pixels and 1.4 μm pixel width.

The calibration field used in this study is a satellite antenna having 150 cm diameter with 112 control points on it. It was chosen since it has a smooth digital surface model and it is geometrically similar to the lens surface model. In this way, the analysis of the errors caused by the geometry of the objective and a balanced distribution of the depth differences over the image acquisition line on the surface model is accomplished. Point location accuracy is approximately 30–35 μm. In order to get the determined point location accuracy, a geodesic Wild T3 theodolite was chosen for direction measurements. In total, five serial measurements were

	iPhone 4S	Nikon Coolpix 8700
Sensor	CMOS 1/3.2″ sensor size	CCD 2/3″
Image resolution	3264 × 2448 (8.0 MP)	3264 × 2448 (8.0 MP)
Focal length*	4.324602 mm	9.027620 mm
Pixel size*	1.4 μm	2.7 μm
Digital zoom values	Up to 5×	Up to 4×
Aspect ratio	4:3	4:3; 3:2
LCD size	3.5″	1.8″

Notes: *These are the values obtained after separate calibrations performed before the application in PI3000 software for iPhone 4S and Nikon Coolpix8700 cameras used in the application. The pictures in the resolution of 3264 × 2448 are taken in this focal length.

Table 1. Technical specifications of the cameras.

made horizontally and vertically with Wild T3 [86]. Used calibration field is complying with self-calibration model, which is a dish antenna model. **Figure 7** shows the images of the calibration field taken by the two camera-lens combinations. The images of the calibration field were taken with the minimum focal length of the each camera without zooming.

The application benefits from the comparison made over iPhone 4S Olloclip 3 in one camera fisheye lens combination with Coolpix8700 camera FC-09 fisheye lens combination in terms of equidistant fisheye model [Eq. (7)] which gives the best result in bundle block adjustment. In the application, the calibration values derived from equidistant fisheye model for iPhone 4S Olloclip 3 in one camera fisheye lens combination and Coolpix8700 camera FC-09 fisheye lens combination are compared. Eqs. (8) and (9) use A_1, A_2, A_3, B_1, B_2, C_1, C_2 coefficients as calibration parameters. Nine of 112 control points are considered passing points, while 103 of them are full control points. Thirteen images of calibration field were taken by iPhone 4S Olloclip 3 in one

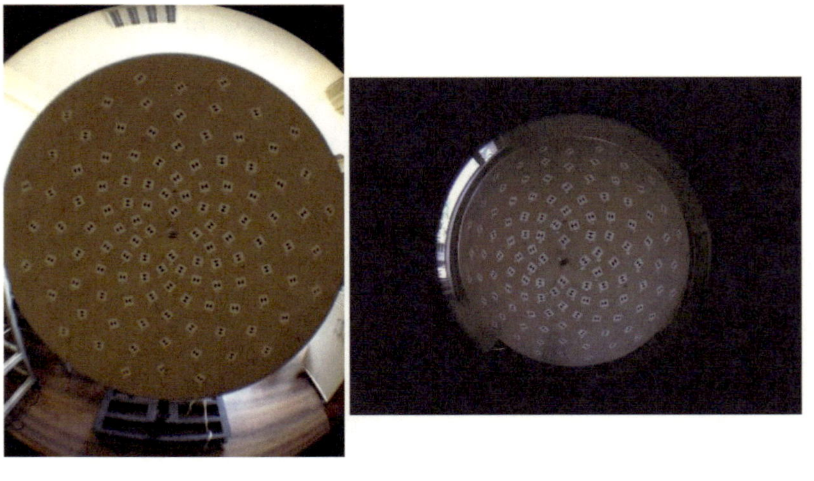

Figure 7. Image of calibration field (Left picture: iphone 4S camera with Olloclip 3 in one; right picture: Nikon Coolpix8700 camera with Nikon FC-E9 fisheye lens).

camera fisheye lens combination from different locations taking into consideration free network adjustment rules. The same procedure was also applied for Coolpix8700 camera FC-09 fisheye lens combination. After then, the two-dimensional image co-ordinates of 112 control points for 13 images were measured in Pictran D software for iPhone 4S Olloclip 3 in one camera fisheye lens combination. The same procedure was also applied for Coolpix8700 camera FC-09 fisheye lens combination. The measurements of the two-dimensional image co-ordinates in Pictran D software are shown in **Figure 8**.

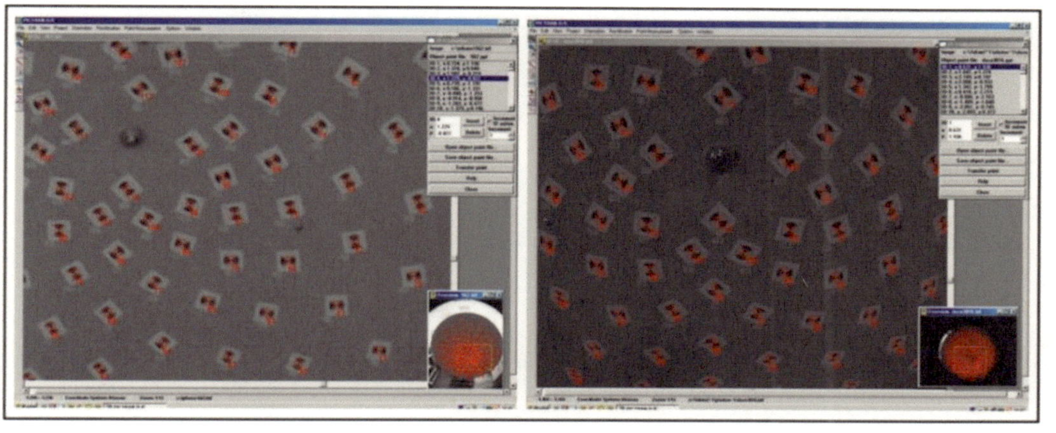

Figure 8. Measurement of image co-ordinates of the fisheye lenses in Pictran D software (Left picture: taken from the combination of iPhone 4S camera with Olloclip 3 in one; right picture: taken from the combination of Nikon Coolpix8700 camera with Nikon FC-E9 fisheye lens).

The resulting image co-ordinates were evaluated in bundle block adjustment software developed by Dr. Danilo Schneider in Dresden Technical University in Germany. According to the bundle block adjustment results, **Tables 2** and **3** were acquired.

The results of bundle block adjustment calculation by the software for 13 images captured with iPhone 4S camera and Olloclip 3 in one are given, respectively; the sigma0 value is 0.00099 and the pixel size is 0.0022 mm. The results of bundle block adjustment calculations by the software for 13 images captured with Nikon Coolpix8700 camera and FC-09 fisheye lens are given, respectively; the sigma0 value is 0.00163 and the pixel size is 0.0022 mm.

	iPhone 4S with Olloclip 3 in one	NIKON Coolpix8700 with NIKON FC-E9
Sigma0	0.00099	0.00163
Convergence	0	0
Max. iteration	100	100
Required iteration	18	32
Calculation time	3.57 sn	6.41 sn
Unknown	396	396
Observations	2868	2838

Table 2. Calibration results of two different fisheye in bundle adjustment software.

	iPhone 4S with Olloclip 3 in one (mm)			Nikon Coolpix8700 with Nikon FC-E9 (mm)		
	Value	**rms**	**Significance**	**Value**	**rms**	**Significance**
c_k	−2.23950000	0.00161000	8	−1.69005000	0.00203000	8
x_0	0.05886000	0.00397000	8	0.03987000	0.00259000	8
y_0	0.10471000	0.00399000	8	0.11442000	0.00240000	8
A_1	0.00080228	0.00027401	8	0.00118750	0.00038919	6
A_2	−0.00386240	0.00007329	8	−0.00067814	0.00005062	8
B_1	−0.00105490	0.00016903	8	−0.00026227	0.00012548	4
B_2	−0.00137450	0.00015347	8	0.00021274	0.00011927	3
C_1	0.00050562	0.00032783	1	−0.00009505	0.00029182	1
C_2	−0.00038251	0.00030980	6	0.00031644	0.00028458	1

Table 3. Calibration parameters calculated for equidistant model.

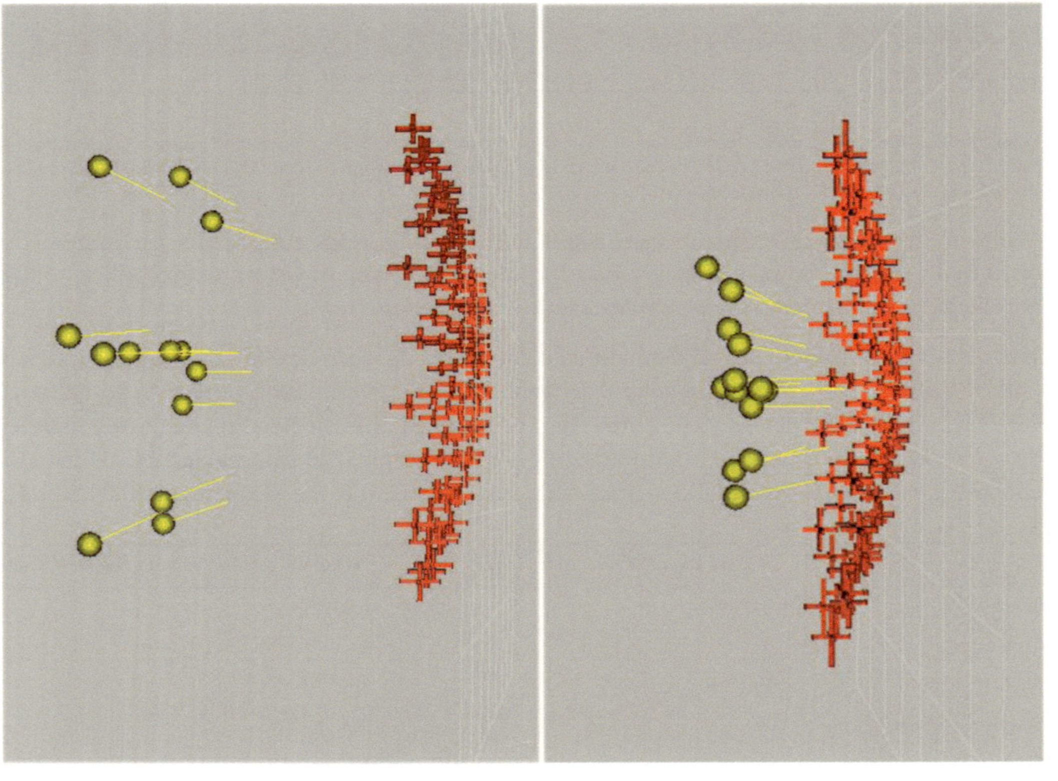

Figure 9. Three-dimensional position of projection points with regard to antenna; three-dimensional co-ordinates of the camera as the result of balancing (left image shows values for iPhone 4S camera with Olloclip 3 in one; and right image shows values for Nikon Coolpix8700 camera with Nikon FC-E9 fisheye lens).

Sigma0 is the root mean square (rms) of measurements for image co-ordinates after bundle block adjustment. **Table 2** shows the resulting values in both of the calibrations. According to software's post-adjustment outputs, additionally, projection point parameters (X_0, Y_0, Z_0 and omega, phi, kappa) of the 13 images and object points' significance for both of the camera and lens combination are 8, which is 99.9%. (Software' significance values are 1: no significance, 2: 80%, 3: 90%, 4: 95%, 5: 98%, 6: 99%, 7: 99.8%, 8: 99.9% high significance) [87].

Figure 9 shows actual positions of three-dimensional co-ordinates of calibration field obtained by adjustment results and projection points of each of the 13 images that come from both of the calibration files acquired after adjustment.

5. Results

At the end of the application designed for testing, numerical values of calibration parameters and rms of those parameters that were calculated according to equidistant model were compared between the 'Olloclip 3 in one' fisheye lens used with iPhone 4S mobile phone and standard 'Nikon FC-E9' fisheye lens used in Nikon Coolpix8700. This comparison is given in **Table 3** and the resulting graphics are shown in **Figures 10** and **11**.

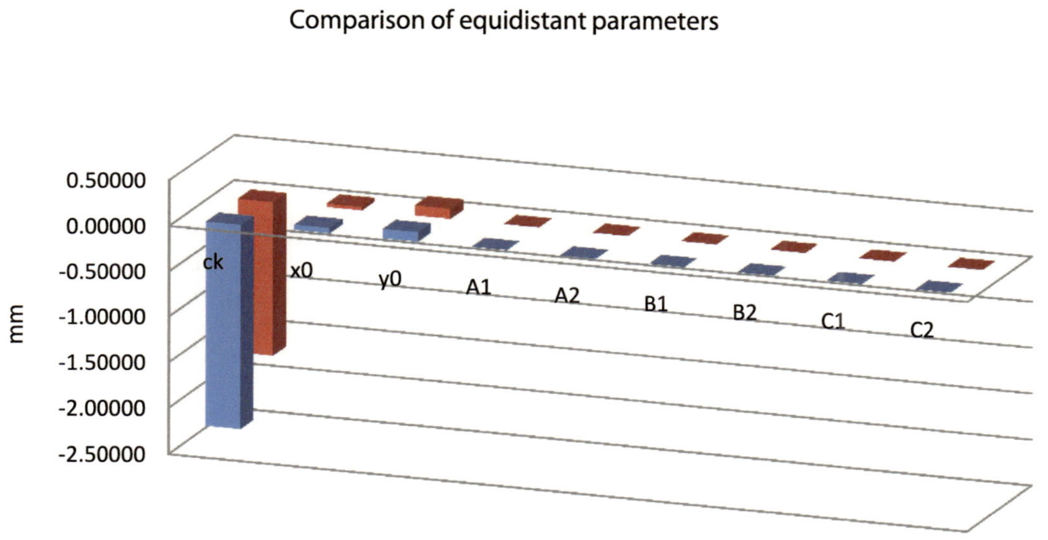

	ck	x0	y0	A1	A2	B1	B2	C1	C2
Olloclip 3 in one	-2.23950	0.05886	0.10471	0.00080	-0.00386	-0.00105	-0.00137	0.00051	-0.00038
Nikon FC-E9	-1.69005	0.03987	0.11442	0.00119	-0.00068	-0.00026	0.00021	-0.00010	0.00032

Figure 10. Distortion parameters for two different camera-fisheye lens combinations.

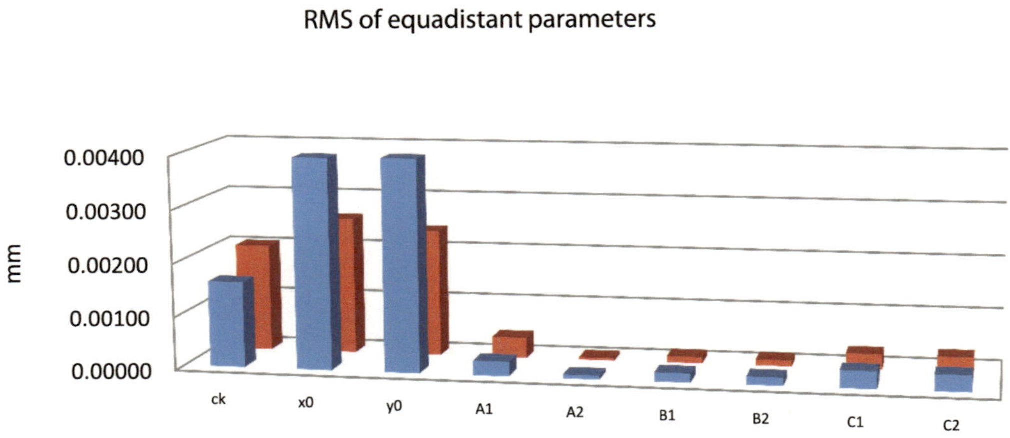

	ck	x0	y0	A1	A2	B1	B2	C1	C2
Olloclip 3 in one	0.00161	0.00397	0.00399	0.00027	0.00007	0.00017	0.00015	0.00033	0.00031
Nikon FC-E9	0.00203	0.00259	0.00240	0.00039	0.00005	0.00013	0.00012	0.00029	0.00028

Figure 11. Rms values of distortion parameters for two different camera-fisheye lens combinations.

Since A_3 distorsion parameter was a so small value that can be ignored, it was not analysed and written in **Table 3** [87]. **Table 3** shows that the significance values of the iPhone are higher than that of Nikon because of smaller pixel (it is given in **Table 1** in Section 3) structure of iPhone's camera.

When **Figure 11** is examined, it is seen that distortion parameters of iPhone 4S camera and Olloclip 3 in one fisheye lens equipment is larger, although it has a larger focal length. It does not show a significant difference than Nikon Coolpix8700 camera and FC-09 fisheye lens equipment. (c_k: focal length after calibration process, x_0: image co-ordinate of principle point in X direction, y_0: image co-ordinate of principle point in Y direction).

Figures 10 and **11** illustrate that when x_0, y_0 principal image point co-ordinate values and focal lengths of two different fisheye lenses are ignored for the cameras with same resolution and same pixel size, a meaningful approximation is obtained. In consideration of these results, current technology developed for Olloclip 3 in one lens that is improved for mobile phones, is particularly great. Nikon FC-E9 mounted on bulky Nikon Coolpix8700 is difficult to use. Olloclip 3 in one lens mounted on iPhone 4S is considered to be used in studies done with photogrammetric fisheye lens instead of Nikon FC-E9. The conclusion part compares the advantages and disadvantages of two different fisheye images.

Nine of 112 control points are considered as passing points, while 103 of them are full control points. One hundred and three point co-ordinates from testing area are considered errorless and used in bundle block adjustment. Three-dimensional position data derived from the bundle block adjustment are compared to the errorless points. Nikon Coolpix8700 camera and FC-09 fisheye lens combination give accuracy on 85 points under the sub-pixel level. IPhone 4S camera and Olloclip 3 in one fisheye lens combination gives accuracy on 89 points under the sub-pixel level. This means that there is 82.52% accuracy for Nikon Coolpix8700 camera and FC-09 fisheye lens combination and 86.40% accuracy for iPhone 4S camera and Olloclip 3 in one fisheye lens combination. The co-ordinates of three-dimensional object co-ordinates of 103 points are subtracted from the points that derived from bundle block adjustment process. If the difference values are greater than sub-pixel level in any axis then they are eliminated consequently. **Figures 12** and **13** are depicted from the obtained differences, respectively, for iPhone 4S camera and Olloclip 3 in one fisheye lens combination and Nikon Coolpix8700 camera and FC-09 fisheye lens combination. Delta X is the difference between three-dimensional point co-ordinate that measured before adjustment in X direction and three-dimensional point co-ordinate, which derived after adjustment in X direction. Delta Y and delta Z were calculated similarly.

The standard deviations of co-ordinate differences have been calculated for three different axes from the data contributing to depict **Figures 12** and **13**. The standard deviation values on the X, Y and Z axes are 0.763, 0.558 and 0.638 mm, respectively, for Olloclip 3 in one lens kit. By using similar derivation, the standard deviation values on the X, Y and Z axes are 0.748, 0.699 and 0.517 mm, respectively, for Nikon FC-09 lens kit. As presented, the standard deviation values of identical axes are found approximately close to each other from the calculations. Moreover, when the distribution of the co-ordinate differences were evaluated for three identical axes (X, Y, Z) of these two kinds of lens kits, it was calculated that they have the same maximum difference value which is approximately 2 mm. From these graphics, root mean square error of point positions has

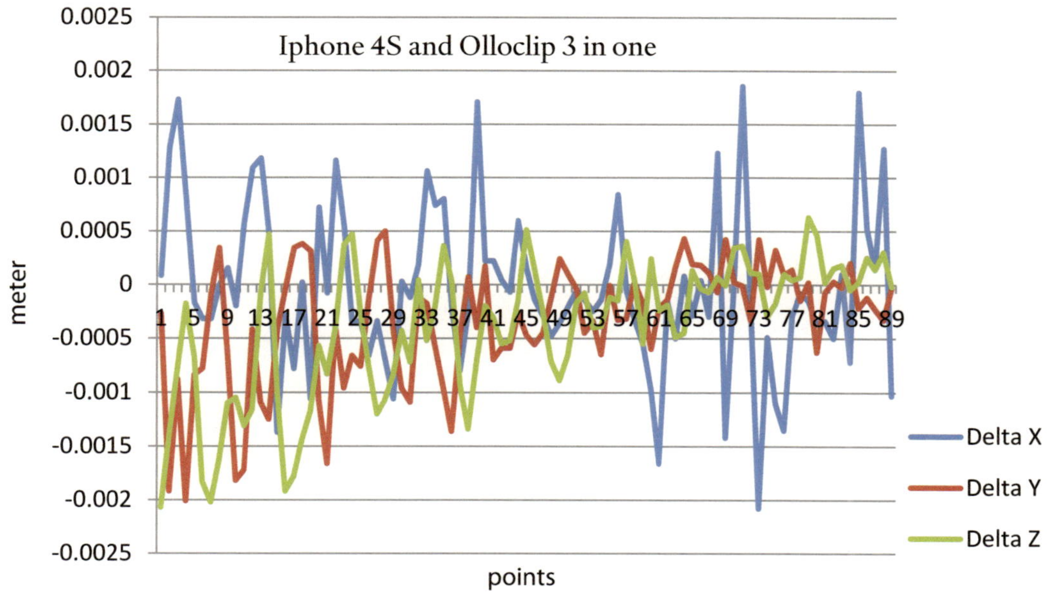

Figure 12. Subpixel graphic for the combination of iPhone 4S camera with Olloclip 3 in one fisheye lens.

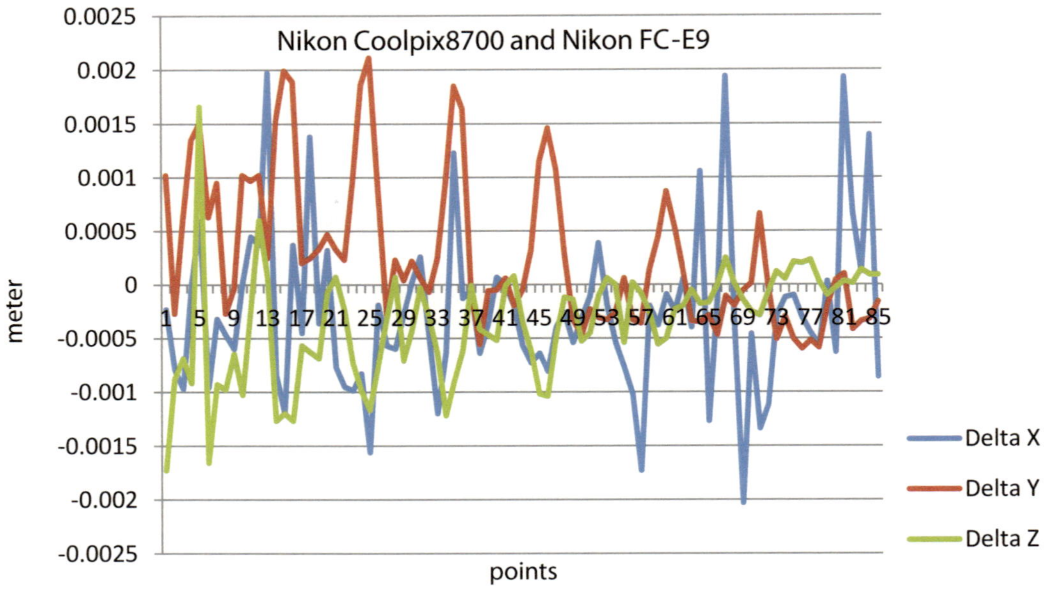

Figure 13. Subpixel graphic for the combination of Nikon Coolpix8700 camera with Nikon FC-E9 fisheye lens.

been determined as 3.556 mm for Olloclip 3 in one and 3.401 mm for Nikon FC-09 lens kit. These values show us the internal reliability of these two kinds of fisheye lens kits is similar for three-dimensional point co-ordinate determination.

When the difference of image co-ordinates derived before and after adjustment are analysed in vectorial form, **Figures 14** and **15** are achieved for the two different fisheye lenses. If the difference

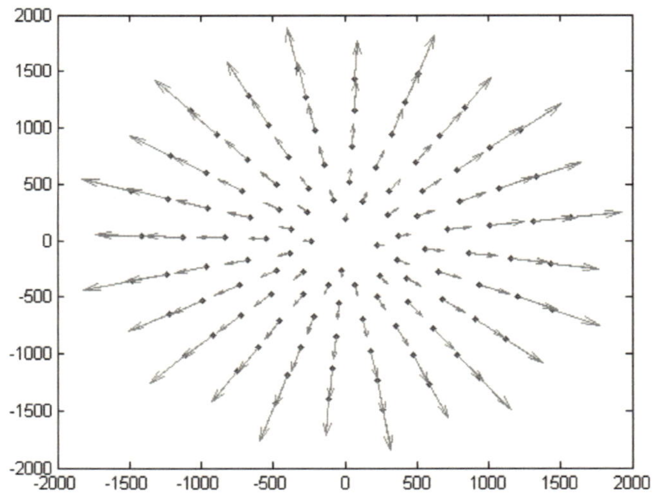

Figure 14. Image co-ordinates residuals for iPhone 4S camera and Olloclip 3 in one fisheye lens combination (between measured and after adjustment).

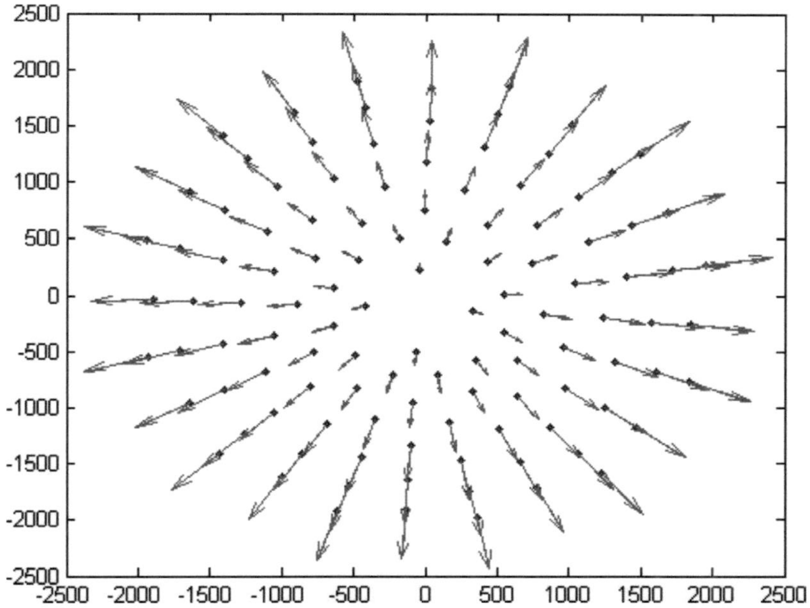

Figure 15. Image co-ordinates residuals for Nikon Coolpix8700 camera and FC-09 fisheye lens combination (between measured and after adjustment).

between the measured images co-ordinates and the adjusted image co-ordinates are analysed in vectorial form, **Figures 14** and **15** are achieved for the two different fisheye lenses. As can be seen from these two figures, residuals decrease while approaching to the image principal point for the two fisheye lens cameras, but increase proportionally to the sides due to distortion. The standard deviation results of the above-mentioned experiment are 0.000814 for iPhone 4S camera and Olloclip 3 in one fisheye lens combination and 0.000890 for Nikon Coolpix8700 camera and FC-09 fisheye lens combination. It can be said that according to above explained results, although there is not a significant difference, iPhone 4S camera and Olloclip 3 in one fisheye lens combination has lower distortion on the total image surface than Nikon Coolpix8700 camera and FC-09 fisheye lens combination. Therefore, iPhone 4S camera and Olloclip 3 in one fisheye lens combination could also be used for photogrammetric applications instead of Nikon Coolpix8700 camera and FC-09 fisheye lens combination.

Advantages and disadvantages of using fisheye lenses for the above-mentioned equipments can be listed as follows:

Advantages

- IPhone 4S camera and Olloclip 3 in one fisheye lens equipment is lightweight and it is much more easy to use.
- As given in **Table 3** in Section 3, focal distance of iPhone 4S camera and Olloclip 3 in one fisheye lens equipment is larger than focal distance of Nikon Coolpix8700 camera and FC-09 fisheye lens equipment. The rms of the larger focal distance is smaller than the other one.

- There is no significant difference between the image centre point co-ordinates of iPhone 4S camera and Olloclip 3 in one fisheye lens equipment and Nikon Coolpix8700 camera and FC-09 fisheye lens.

Disadvantages

- Nikon Coolpix8700 camera and FC-09 fisheye lens equipment are heavier and much more difficult to use.
- Since focal distance of iPhone 4S camera and Olloclip 3 in one fisheye equipment is larger than Nikon Coolpix8700 camera and FC-09 fisheye lens equipment, the resulting distortion parameters are expected to be smaller and when distortion parameters of both equipments are compared to each other, there happens to be a stable result that exceeds the expectations.
- As given in **Table 3** in Section 3, mean square error values for image central point of 4S camera and Olloclip 3 in one fisheye lens equipment are higher than mean square error values of Nikon Coolpix8700 camera and FC-09 fisheye lens equipment. Therefore, Nikon Coolpix8700 camera and FC-09 fisheye lens equipment can be considered to be more stable.
- IPhone 4S camera and Olloclip 3 in one fisheye lens equipment should be tested in a photogrammetric study and the results should be interpreted in the light of these data.

6. Conclusion

The main purpose of this study is to test fisheye lens equipment used with mobile phones. In this study, the performance of Olloclip 3 in one fisheye lens used with iPhone 4S mobile phone and Nikon FC-E9 fisheye lens used with Nikon Coolpix8700 camera is analysed comparing the calibration results based on an equidistant model. The resolution of the cameras is the same for these two kinds of hardware. The co-ordinates of image centre point were found approximately close to each other from the calculations for these two kinds of hardware. It was seen that the calibration results of Olloclip 3 in one fisheye lens used with iPhone 4S mobile phone have not showed statistical significant difference results compared to Nikon FC-E9 fisheye lens used with Nikon Coolpix8700. In addition, it was seen from the results of this study that Olloclip 3 in one fisheye lens has larger focal length than the other. This experimental study shows that Olloclip 3 in one fisheye lens developed for mobile phones has at least the similar characteristics with classic fisheye lenses.

Smartphones and fish eye lens are very popular devices in developing computer technologies. The use of fisheye lenses having big distortion was limited in the past, but today use of advanced computer software ease the solution of distortion problem. Therefore, fisheye lenses became the mostly studied devices and issues in photogrammetric fieldwork. Additionally, the use of fisheye lenses together with smartphones has opened new research areas. The dimensions of fisheye lenses used with smartphones are getting smaller and the prices are reducing. Moreover, as verified in this study, the accuracy of fisheye lenses used in smartphones is better than conventional fisheye lenses. The use of smartphones with fisheye lenses will give the possibility of practical applications to ordinary users in the near future.

Author Biography

Cumhur Sahin was born in 1977. He got his bachelor's degree in 2001 in Istanbul Technical University in Turkey. He completed his post-graduation in 2004 with Master of Science in Geodesy and Photogrammetry Engineering in Gebze Institute of Technology in Turkey. He obtained his doctorate in 2011 in Geodesy and Photogrammetry Engineering in Yildiz Technical University in Turkey. He worked as a research assistant in Gebze Institute of Technology from 2001 to 2014 in Turkey. He has been working as a research assistant in Gebze Technical University in Turkey since 2014.

Author details

Cumhur Sahin

Address all correspondence to: csahin@gtu.edu.tr

Department of Geodetic and Photogrammetric Engineering, Faculty of Engineering, Gebze Technical University, Gebze, Kocaeli, Turkey

References

[1] iPhone Lens Kits Reviews [Internet]. 2015. Available from: http://iphone-lens-kits-review.toptenreviews.com/ [Accessed: April 21, 2017]

[2] Available from: http://www.mobilephonecameralenses.com/ [Accessed: April 21, 2017]

[3] Olloclip [Internet]. 2015. Available from: https://www.olloclip.com/ [Accessed: April 21, 2017]

[4] Chugh G, Bansal D, Sofat S. Road condition detection using smartphone sensors: A survey. International Journal of Electronic and Electrical Engineering. 2014;7(6):595–602. ISSN 0974-2174

[5] Perttunen M, Mazhelis O, Cong F, Kauppila M, Leppanen T, Kantola J, Collin J, Pirttikangas S, Haverinen J, Ristaniemi T, Riekki J. Distributed road surface condition monitoring using mobile phones. In: Proceeding of 8th International Conference UIC 2011; September 2–4, 2011; Banff, Canada. ISBN: 978-3-642-23640-2. Springer, Berlin Heidelberg, Germany, 2011. doi:10.1007/978-3-642-23641-9_8

[6] Mohan P, Padmanabhan VN, Ramjee R. Nericell: Rich monitoring of road and traffic conditions using mobile smartphones. In: Proceeding of the 6th ACM conference on Embedded network sensor systems (SenSys'08); November 5–7, 2008; Raleigh, North Carolina, USA.

[7] Wagner D, Reitmayr G, Mulloni A, Drummond T, Schmalstieg D. Pose tracking from natural features on mobile phones. In: Proceeding of IEEE International Symposium on Mixed and Augmented Reality; September 15–18, 2008; Cambridge, UK

[8] Prisacariu VA, Kahler O, Murray D, Reid I, Simultaneous 3D tracking and reconstruction on a mobile phone. In: Proceeding of IEEE International Symposium on Mixed and Augmented Reality (ISMAR); October 1–4, 2013; Adelaide, SA, Australia. p. 89–98.

[9] Tanskanen P, Kolev K, Meier L, Camposeco F, Saurer O, Pollefeys M. Live Metric 3D reconstruction on mobile phones. In: Proceeding of IEEE International Conference on Computer Vision (ICCV); December 1–8, 2013; Sydney, NSW, Australia. pp. 55–62

[10] Pan Q, Arth C, Reitmayr G, Rosten E, Drummond T. Rapid scene reconstruction on mobile phones from panoramic images. In: Proceeding of IEEE International Symposium on Mixed and Augmented Reality; October 26–29, 2011; Basel, Switzerland. 2011. pp. 55–64

[11] Wagner D, Mulloni A, Langlotz T, Schmalstieg D. Real-time panoramic mapping and tracking on mobile phones. In: Proceeding of IEEE International Symposium on Virtual Reality; March 20–24, 2010; Waltham, Massachusetts, USA. pp. 211–218

[12] Almazan J, Bergasa LM, Yebes JJ, Barea R, Arroyo R. Full auto-calibration of a smartphone on board a vehicle using IMU and GPS embedded sensors. In: Proceeding of IEEE International Symposium on Intelligent Vehicles Symposium (IV); June 23–26, 2013; Gold Coast, Australia. pp. 1374–1380

[13] Hill A, MacIntyre B, Gandy M, Davidson B, Rouzati H. KHARMA: An Open KML/HTML Architecture for mobile augmented reality applications. In: Proceeding of IEEE International Symposium on Mixed and Augmented Reality; October 13–16, 2010; Seoul, Korea. pp. 233–234

[14] Kurz D, Benhimane S. Gravity-aware handheld augmented reality. In: Proceeding of IEEE International Symposium on Mixed and Augmented Reality; October 26–29, 2011; Basel, Switzerland. pp. 111–120

[15] Porzi L, Ricci E, Ciarfuglia TA, Zanin M. Visual-inertial tracking on android for augmented reality applications. In: Proceeding of IEEE International Symposium on Environmental Energy and Structural Monitoring Systems (EESMS); September 28, 2012; Perugia, Italy

[16] Schindhelm CK, Gschwandtner F, Banholzer M. Usability of apple iPhones for inertial navigation systems. In: Proceeding of IEEE International Symposium on 22nd International Symposium on Personal, Indoor and Mobile Radio Communications; September 11–14, 2011; Toronto, ON, Canada. pp. 1254–1258

[17] Burgess S, Åström K, Lindquist B, Ljungberg R, Sharma K. Indoor localization using smartphones in multi floor environments without prior calibration or added infrastructure. In: Proceeding of International Conference on Indoor Positioning and Indoor Navigation (IPIN); October 13–16, 2015; Banff, Alberta, Canada

[18] Burgess S, Åström K, Högström M, Lindquist B, Ljungberg R. Smartphone positioning in multi-floor environments without calibration or added infrastructure. In: Proceeding of IEEE International Conference on Indoor Positioning and Indoor Navigation, (IPIN); October 4–6, 2016; Madrid, Spain.

[19] Hadid A, Heikkila JY, Silven O, Pietikainen M. Face and eye detection for person authentication in mobile phones. In: Proceeding of IEEE International Symposium on Distributed Smart Cameras, ICDSC '07; September 25–28, 2007; Vienna, Austria.

[20] Shen Y, Hu W, Yang M, Wei B, Luce S, Chou CT. Face recognition on smartphones via optimised sparse representation classification. In: Proceeding of IEEE International Conference on Information Processing in Sensor Networks, IPSN'14; April 15–17, 2014; Berlin, German. p. 237–248.

[21] Wang Y, Tan R, Xing G, Wang J, Tan X, Liu X. Energy-efficient aquatic environment monitoring using smartphone-based robots. ACM Transactions on Sensor Networks. 2016;**12**(3):25. DOI: 10.1145/2932190

[22] Sreelekshmi B, Vidya N. Survey on monitoring aquatic debris using Smartphone-Based robots. International Journal of Engineering and Computer Science. 2015;**58**(10):20131–20133. DOI: 10.18535/ijecs/v6i1.50

[23] Maindalkar AA, Ansari SM. Aquatic robot design for water pollutants monitoring. International Journal on Recent and Innovation Trends in Computing and Communication. 2015;**3**(6):3699–3703. ISSN: 2321-8169

[24] Muaremi A, Arnrich B, Troster G. Towards Measuring stress with smartphones and wearable devices during workday and sleep. BioNanoScience. 2013;**3**:172–183. DOI 10.1007/s12668-013-0089-2

[25] Miluzzo E, Wang T, Campbell AT. EyePhone: Activating mobile phones with your eyes. In: Proceeding of the 2nd ACM SIGCOMM Workshop on Networking, Systems, and Applications for Mobile Handhelds, MobiHeld; August 30, 2010; New Delhi, India. DOI: 10.1145/1851322.1851328

[26] Sandstrom GM, Lathia N, Mascolo C, Rentfrow PJ. Putting mood in context: Using smartphones to examine how people feel in different locations. Journal of Research in Personality. 6 pages. Open access, Available online 8 June 2016. http://dx.doi.org/10.1016/j.jrp.2016.06.004

[27] Lane ND, Xu Y, Lu H, Hu S, Choudhury T, Campbell AT, Zhao F. Enabling large-scale human activity Inference on smartphones using community similarity networks (CSN). In: Proceeding of 13th ACM International Conference Ubiquitous Computing, UbiComp11; September 17–21, 2011; Beijing, China. p. 355–364. DOI: 10.1145/2030112.2030160.

[28] Lu H, Yang J, Liu Z, Lane ND, Choudhury T, Campbell AT. The Jigsaw continuous sensing engine for mobile phone applications. In: Proceeding of 8th ACM International Conference on Embedded Networked Sensor Systems, SenSys'10; November 3–5, 2010; Zurich, Switzerland. DOI: 10.1145/1869983.1869992.

[29] Lu H, Rabbi M, Chittaranjan GT, Frauendorfer D, Mast MS, Campbell AT, Perez DG, Choudhury T. StressSense: detecting stress in unconstrained acoustic environments using smartphones. In: Proceeding of 12th ACM International Conference Ubiquitous Computing, UbiComp '12; September 5-Sep 8, 2012; Pittsburgh, USA

[30] Lane ND, Miluzzo E, Lu H, Peebles D, Choudhury T, Campbell AT, College D. A Survey of mobile phone sensing. IEEE Communications Magazine, 2010;**48**(9):140-150. DOI: 10.1109/MCOM.2010.5560598.

[31] Rachuri KR, Musolesi M, Mascolo C, Rentfrow P, Longworth C, Aucinas A. EmotionSense: A mobile phones based adaptive platform for experimental social psychology research. In: Proceeding of 10th ACM International Conference Ubiquitous Computing, UbiComp '10; September 26-29, 2010; Copenhagen, Denmark. DOI: 10.1145/1864349.1864393

[32] LiKamWa R, Liu Y, Lane ND, Zhong L. Can your smartphone infer your mood? In: Proceeding of Second International Workshop on Sensing Applications on Mobile Phones at ACM, SenSys '11; November 1, 2011; Seattle, USA

[33] Ha H, Bok Y, Joo K, Jung J, Kweon IS. Accurate camera calibration robust to defocus using a smartphone. This ICCV paper is the Open Access version. This ICCV paper is the Open Access version, provided by the Computer Vision Foundation. Except for this watermark, it is identical to the version available on IEEE Xplore. In: IEEE International Conference on Computer Vision (ICCV). 2015. pp. 828–836. DOI: 10.1109/ICCV.2015.101

[34] Gruen A, Akca D. Calibration and accuracy testing of mobile phone cameras. In: Proceeding of the 28th Asian Conference on Remote Sensing, ACRS '07; November 12–16, 2007; Kuala Lumpur, Malaysia. pp. 1171–1177

[35] Delaunoy A, Li J, Jacquet B, Pollefeys M. Two Cameras and a Screen: How to Calibrate Mobile Devices? In: Proceeding of 2nd IEEE International Conference on 3D Vision; December 8–11, 2014; Tokyo, Japon. p.123–130

[36] Saponaro S, Kambhamettu C. Towards auto-calibration of smart phones using orientation sensors. In: Proceeding of IEEE Conference on Computer Vision and Pattern Recognition Workshops. June 23–28, 2013; Portland, Oregon, USA. pp. 20–26.

[37] Ahn H, Choi C, Yeu Y. Assessment of smartphone camera in close-range digital photogrammetry. In: Proceeding of 34th Asian Conference on Remote Sensing, ACRS '13; October 20–24, 2013; Bali, Indonesia. pp. 1630–1634

[38] Abraham S, Forstner W. Fish-eye-stereo calibration and epipolar rectification. ISPRS Journal of Photogrammetry and Remote Sensing. 2005;**59**(2):278–288

[39] Gledhill D, Tian GT, Taylor D, Clarke D. Panoramic imaging—a review. Computers & Graphics. 2003;**27**(3):435–445

[40] Schwalbe, E. Geometric modelling and calibration of fisheye lens camera systems. In: Proceedings of the 2nd ISPRS Panoramic Photogrammetry Workshop; February 24–25, 2005; Berlin, Germany. p. 5–8.

[41] Kumler J, Bauer M. Fisheye lens designs and their relative performance. In: Proceedings of the Current Developments in Lens Design and Optical Systems Engineering, Vol. 4093. SPIE '00; October 24, 2000; San Diego, Calif, USA. pp. 360–369

[42] Dhane P, Kutty K, Bangadkar S. A generic non-linear method for fisheye correction. International Journal of Computer Applications. 2012;**51**(10):58–65

[43] Yamashita M, Yoshimura M, Nakashizuka T. Cloud cover estimation using multitemporal hemisphere imageries. In: Proceedings of the Photogrammetry and Remote Sensing, 20th ISPRS Congress, ISPRS '04; July 12–23, 2004; Istanbul, Turkey. pp. 826–829

[44] Wang L, Duan F, Lv K. Fisheye-lens-based visual sun compass for perception of spatial orientation. Mathematical Problems in Engineering, 2012;**2012**(460430):15pages. http://dx.doi.org/10.1155/2012/460430.

[45] Oh JKW, Haberl JS. New educational software for teaching the sunpath diagram and shading mask protractor. In: Proceedings of Conference on the 5th International Building Performance Simulation Association, IBPSA; September 8–10, 1997; Prague, Czech Republic. 7

[46] Beekmans C, Schneider J, Läbe T, Lennefer M, Stachniss C, Simmer C. Cloud photogrammetry with dense stereo for fisheye cameras. Atmospheric Chemistry and Physics. 2016;**16**:14231–14248. DOI: 10.5194/acp-16-14231-2016

[47] Krishnan G, Nayar SK. Cata-fisheye camera for panoramic imaging. In: Proceedings of the IEEE Workshop on Applications of Computer Vision; January 7–8, 2008; Copper Mountain, Colo, USA. pp. 1–8.

[48] Streckel B, Evers-Senne JF, Koch R. Lens model selection for a markerless AR tracking system. In: Proceedings of the 4th IEEE and ACM International Symposium on Mixed and Augmented Reality, October 22–25, 2005; Vienna, Austria. pp. 130–133

[49] Brun X, Deschaud JE, Goulette F. On-the-way city mobile mapping using laser range scanner and fisheye camera. In: Proceedings of the 5th ISPRS International Symposium on Mobile Mapping Technology, MMT '07; May 29–31, 2007; Padua, Italy, 8 pages.

[50] Yamamoto D, Ozeki S, Takahashi N. Wired fisheye lens: a motion-based improved fisheye interface for mobile web map services. In Web and Wireless Geographical Information Systems. Springer, ISSN: 0302-9743. In: Proceedings of 9th International Symposium, W2GIS 2009; December 7–8, 2009; Maynooth, Ireland.

[51] Ahmad A, Lima P. Multi-robot cooperative spherical object tracking in 3D space based on particle filters. Robotics and Autonomous Systems, 2013;**61**(10):1084–1093

[52] Zheng JY, Li S. Employing a fish-eye for scene tunnel scanning. In: Proceedings of the 7th Asian Conference on Computer Vision, ACCV '06; January 13–16, 2006; Hyderabad, India. pp. 509–518

[53] Martinez E, Del Pobil AP. A dioptric stereo system for robust real-time people tracking. In: Proceeding of IEEE International Conference on Robotics and Automation (ICRA '09); May 12–17, 2009; Kobe, Japan

[54] Gurtner A. Investigsation of fisheye lenses for small UAV aerial photography [thesis]. Brisbane, Australia: Queensland University of Technology; 2008

[55] Grelsson B. Global pose estimation from aerial images registration with elevation models [thesis]. Swedish: Linköping University Electronic Press, Linköpings Universitet, Linköping; 2014.

[56] Naruse T, Kaneko T, Yamashita A, Asama H. 3-D measurement of objects in water using fish-eye stereo camera. In: Proceeding of the 19th IEEE International Conference on Image Processing (ICIP '12); September 30–October 3, 2012; Orlando, Fla, USA. 2012. pp. 2773–2776

[57] Bouaziz S. Geolocalization using skylines from omni-images [thesis]. Lausanne, Switzerland: Virtual Reality Laboratory,'Ecole Polytechnique F'ed'erale de Lausanne; 2009

[58] Georgantas A, Br'edif M, Desseilligny MP. An accuracy assessment of automated photogrammetric techniques for 3D modelling of complex interiors. In: Proceeding of 22th the ISPRS Congress. Vol. 39-B3. August 25–September 1, 2012; Melbourne, Australia. 2012:23–28

[59] Stehle T, Truhn D, Aach T, Trautwein C, Tischendorf J. Camera calibration for fish-eye lenses in endoscopywith an application to 3Dd reconstruction. In: Proceedings of IEEE International Symposium on Biomedical Imaging (ISBI); April 12–15, 2007; Washington, D.C. pp. 1176–1179

[60] Hu P, Shen G, Li L, Lu D. ViRi: View it Right. In: Proceedings of the 11th ACM Annual International Conference on Mobile Systems, Applications, and Services, MobiSys'13; June 25–28, 2013; Taipei, Taiwan.

[61] Schneider D, Schwalbe E. Integrated processing of terrestrial laser scanner data and fisheye-camera image data. In: Proceedings of the 21st ISPRS Congres, Vol. XXXVII Part B5; July 3–11, 2008; Beijing, China. p. 1037–1044

[62] Hughes C, Denny P, Jones E, Glavin M. Accuracy of fisheye lens models. Applied Optics. 2010;**49**(17):33383347

[63] Junior JM, Moraes MVA, Tommaselli AMG. Experimental assessment of techniques for fisheye camera calibration. Boletim de Ciências Geod'esicas. 2015;**21**(3):637–651

[64] Arfaoui A, Thibault S. Fisheye lens calibration using virtual grid. Applied Optics. 2013; **52**(12):2577–2583

[65] Kim D, Paik J. Three-dimensional simulation method of fish-eye lens distortion for a vehicle backup rear-view camera. Journal of the Optical Society of America A, 2015;**32**(7): 1337–1343

[66] Torii A, Havlena M, Pajdla T. Omnidirectional image stabilization for visual object recognition. International Journal of Computer Vision. 2011;**91**:157–174. DOI: 10.1007/s11263-010-0350-x

[67] Schneider D, Schwalbe E, Maas H-G. Validation of geometric models for fisheye lenses. ISPRS Journal of Photogrammetry and Remote Sensing. 2009;**64**(3):259–266

[68] Kannala J, Brandt SS. A generic camera model and calibration method for conventional, wide-angle, and fish-eye lenses. IEEE Transactions on Pattern Analysis and Machine Intelligence. 2006;**28**(8):1335–1340

[69] Parian JA. Panoramic imaging: Techniques, sensor modeling and applications. In: Proceedings of the International Summer School on Digital Recording and 3D Modelling. April 2006; Crete, Greece

[70] Zhu H, Wang X, Zhou J, Wang X. Approximate model of fisheye camera based on the optical refraction. Multimedia Tools and Applications. 2014;**73**:1445–1457. DOI 10.1007/s11042-013-1641-3

[71] Micusik B, Pajdla T. Estimation of omnidirectional camera model from epipolar geometry. In: Proceeding of IEEE Computer Vision and Pattern Recognition, June 16–22, 2003; Madison, Wisconsin, USA. pp. 485–490

[72] Thirthala S, Pollefeys M. The radial trifocal tensor: A tool for calibrating the radial distortion of wide-angle cameras. In: Proceedings of the IEEE Computer Society Conference on Computer Vision and Pattern Recognition CVPR '05. Vol. 1. June 20–25, 2005; San Diego, California, USA. pp. 321–328

[73] Hughes C, Glavin M, Jones E. Simple fish-eye calibration method with accuracy evaluation. Electronic Letters on Computer Vision and Image Analysis. 2011;**10**(1):54–62

[74] Friel M, Hughes C, Denny P, Jones E, Glavin M. Automatic calibration of fish-eye cameras from automotive video sequences. Intelligent Transport Systems. 2009;**4**(2):136–148. DOI: 10.1049/iet-its.2009.0052

[75] Schmid HH. General Analytical Solution to the Problem of Photogrammetry. Aberdeen: US Ballistic Research Laboratories; 1959. Accession Number: AD0230349.

[76] Mikhail EM, Bethel JS, McGlone JC. Introduction to Modern Photogrammetry. New York: John Wiley & Sons; 2001

[77] Hou W, Ding M, Qin N, Lai X. Digital deformation model for fisheye image rectification. Optics Express. 2012;**20**(20):22252–22261

[78] Johnson KB. Development of a versatile wide-angle lens characterization strategy for use in the OMNIster stereo vision system [thesis]. USA: University of Tennessee, Knoxville, Tenn; 1997

[79] Miyamoto K. Fish eye lens. Journal of the Optical Society of America. 1964;**54**(8):1060–1061

[80] Hughes C, Denny P, Glavin M, Jones E. Equidistant fisheye calibration and rectification by vanishing point extraction. IEEE Transactions on Pattern Analysis and Machine Intelligence. 2010;**32**(12):2289–2296

[81] Hughes C, McFeely R, Denny P, Glavin M, Jones E. Equidistant (fθ) fish-eye perspective with application in distortion centre estimation. Image and Vision Computing. 2010;**28**(3): 538–551

[82] Parhizgar S. An evaluation of an omni-directional (fisheye) vision sensor for estimating the pose of an object [thesis]. Saskatchewan, Canada: University of Regina, Master of Applied Science in Software Systems Engineering University of Regina; 2014

[83] Nikon [Internet]. 2015. Available from: http://imaging.nikon.com/ [Accessed: 21 April 2017]

[84] Image Resource [Internet]. 2015. Available from: http://www.imaging-resource.com/. [Accessed: 2017-04-21]

[85] Photo.net. [Internet]. 2015. Available from: http://photo.net/ [Accessed: 2017-04-21]

[86] Sahin C. A construction of camera calibration field for close range photogrammetric camera calibration [thesis]. Kocaeli, Turkey: Gebze Institute of Technology, Graduate School of Science Engineering and Sciences; 2004

[87] Sahin C. Comparison and calibration of mobile phone fisheye lens and regular fisheye lens via equidistant model. Journal of Sensors. 2016;**2016**(9379203):11 pages. http://dx.doi.org/10.1155/2016/9379203.

Chapter 3

SOS Message Distribution for Searching Disaster Victims

Jane Louie Fresco Zamora, Noriyuki Suzuki and
Shigeru Kashihara

Additional information is available at the end of the chapter

http://dx.doi.org/10.5772/intechopen.69690

Abstract

This chapter presents the design and implementation of the SOSCast application which enables SOS message distribution for searching victims in a disaster-damaged area. During catastrophic disasters such an earthquake or tsunami, people may be injured or trapped in fallen buildings and debris. In situations like these, it is critical that rescue operations must be done within the first 72 h to save many lives. It is also during these events where communication infrastructures are severely damaged, and thus, makes it difficult for victims to ask for help due to the absence of communication channels. By using the SOSCast application in such scenario, victims are able to exchange SOS messages automatically by communicating directly among smartphones with less operation. By collecting these SOS messages, rescuers can find the existences of the victims as mapped on their smartphones. We have shown in our preliminary experiment within a residential area that SOSCast is capable of determining the existence of a propagator based on the collected SOS messages.

Keywords: smartphone, application, emergency communication, SOS message, bluetooth

1. Introduction

In a catastrophic disaster such as the Great Eastern Japan earthquake of 2011, victims of the disaster must be rescued at least within 72 h because the chances of surviving typically decreases thereafter. However, during search and rescue operations, it is difficult to estimate the damage on the affected area. In addition, determining the existences of victims has been proven difficult with the absence of communication channels as they may have been destroyed. In fact, during the 2011 Great Eastern Japan earthquake, 29,000 base stations ceased to operate [1].

The government responded by administering (1) emergency message service, (2) specially installed public telephones, (3) satellite-based mobile phones, (4) mobile telecommunications equipment, (5) amateur radio, and (6) free Internet connection. Nonetheless, it was not easy to dispatch these services immediately after the incident. Also, the rescue teams may not be dispatched right away due to insufficient information.

Thus far, several solutions for maintaining communication in a disaster-affected area with destroyed communication equipment have been proposed in Refs. [2–4]. These typically enable direct communications between mobile devices such as the smartphone via Bluetooth and store-carry-forward (SCF) routing. In recent technology, smartphones have built-in devices like the Bluetooth and WiFi that permits data transmission even without the Internet. Developing software-based applications is one way to take advantage of the features of smartphone devices. With SOSCast that we designed, we aim to find the existence of the victims of disaster in support of the search and rescue operations. As this application make use of the Bluetooth as a communication device, it is possible to establish direct connection between smartphone devices. This then allows victims to send each other SOS messages and at the same time ensure that these messages can be delivered to the rescuers.

In this chapter, we discuss in detail the implementation and design of the SOSCast application that enables Android OS-based smartphones to send SOS messages directly via Bluetooth communication. It must be noted that throughout this chapter, we refer to two kinds of victims. One is the mobile victim who is capable of moving around the damaged area. The other is the immobilized victim, an individual who may be trapped within destroyed buildings or one who is badly injured and incapable of moving away from the current location. Moreover, we refer to rescuers as a group of people authorized to perform the search and rescue operations who come from outside of the disaster area. In general, it is challenging for the rescuers to estimate the location of an individual because the GPS information of the device that the victim has on hand may not be accurate. It is most especially difficult to identify indoor location due to signal blockade from damage buildings and absence of communication infrastructures. However, SOSCast attempts to support the rescuers in locating them most especially the victims who need immediate medical care. With the development of SOSCast, we aim to contribute the following features:

- Estimation of the area where the immobilized persons exist via SOS messages
- Definitions of data format for SOS messages to support searching immobilized persons
- Propagation method for SOS messages directly among smartphones
- Information deletion of rescued immobilized persons

Based on an experiment within a residential area using SOSCast, we present how SOSCast works and how it potentially locate immobilized victims even when the individual is located indoors. The rest of the chapter is organized as follows. Section 2 discusses related work. Section 3 presents the detailed explanation of the application design. Section 4 describes the actual use of the application in an experiment. Finally, Section 5 provides the chapter summary.

2. Related work

In a natural disaster such as a devastating earthquake, typical medium of contacting friends and family nowadays is via a mobile cellular device. However, when damaged areas are most likely left with destroyed communication infrastructures, victims would find it hard to communicate especially for those victims who are immobilized. Several studies have thought about utilizing the mobile cellular device, more commonly known as the smartphone, in sending information even with the absence of communication channels. For example, research by Sakurai et al. [5] proposes a minimalist information system that provides the functions of identifying victims, compiling these information in a database and, matching these information with relief solutions. However, in a worst case scenario, even with limited bandwidth, battery depletion and the guarantee of the information in the database were not considered as we do in SOSCast.

On a similar study that addresses the loss of communication in a disaster-affected area, Sakano et al. [6] suggested the deployment of a "movable and deployable resource unit (MDRU)" to reclaim connectivity. They have proposed a network architecture that utilizes these MDRU to restore the communication channels in a badly damaged area as per lesson from the Great Eastern Japan earthquake of 2011. It is commendable as to how they have demonstrated the effectiveness of the proposed architecture with the MDRUs by simulation but is yet to be tested in actual scenario. However, these MDRUs are estimated to be set up about 2 days and are the size of a typical container van. Thus, it implies that this solution focuses more on the post disaster recovery operations. As with SOSCast, we are more concerned on confirming the existence of victims right after the disaster has occurred and at least within 72 h of estimated survival rate.

Another related study to SOSCast is this proposal of a communication scheme between mobile devices even without communication infrastructures. Nishiyama et al. [7] developed an information relay technique between smartphones that enables the device to switch between MANET and DTN communication mode. The switch allows for the conservation of resource such as battery and bandwidth. In an actual experiment they conducted, they have successfully relayed information from the source to the destination for within 2.5 km even when the source is located indoors. The idea of a message relay via smartphone is attractive since it is important that the message be delivered to the rescuers. However, this study does not generally addresses accounting the victims in a disaster area as it is only concerned with getting the message to the destination. With SOSCast, the application is capable of identifying victims, accumulate these information and effectively deliver these to the rescuers.

Having similiar goals with SOSCast, WIISARD or Wireless Internet System for Medical Response in Disasters aims to establish reliable connections even with the absence of communication channels. Chipara et al. [8] describes WIISARD as an emergency response system having actually tested the system through an emergency drill. It required the deployment of the Calmesh wireless networking node for the Intelligent Triage Tags (ITT) and iMOX device, which monitors the victim's oxygen level, as well as the network of PDA's or tablets with developed software to complete the system. Meanwhile, another emergency response system called DistressNet is similar to that of WIISARD. Collaborating with an actual search and rescue team, Chenji et al. [9] aims to establish a disaster response system using off-the-shelf devices to establish a reliable network of sensors and communication. To do so, routers and vibration sensors need to be deployed by the search and rescue team members to create the mesh network and to successfully triage the disaster victims. In contrast to both emergency response systems, SOSCast does not require the deployment of additional devices or infrastructure to enable triage of immobilized victims. It is sufficient that the victim has a smartphone on hand installed with SOSCast application for the victim to be accounted for.

3. SOSCast application design

The SOSCast application was developed to enable direct communication between victims during a catastrophic disaster particularly when the communication channels are unavailable. It allows for the victims to exchange SOS messages that contains information of their current physical status, GPS location information, etc., via the smartphone using Bluetooth communications. It must be noted that the victims we refer to are both mobile and immobilized victim. A victim is mobile when the individual is capable of traversing the damaged area while at the same time helping out other victims. Immobilized victims on the other hand are individuals who are badly injured or trapped between debris and are unable to move from within their current location. It is assumed that the application will be utilized in a scenario where the disaster-affected area is cut off from the conventional communication services, i.e., landline and cellular. That is, victims having smartphones with the installed application are able to communicate directly. In this section we will discuss in detail how SOSCast functions when it is in use. Section 3.1 describes the general function of the SOSCast. Section 3.2 presents the message format of SOSCast. Section 3.3 describes the propagation process of the SOSCast. Section 3.4 discusses how duplicate messages are avoided to prevent unnecessary confusion. Section 3.5 explains how SOSCast finds the existence of the victims.

3.1. General overview and requirements

Figure 1 shows the general picture of the SOSCast message exchange. The roles, as defined by how the SOSCast application is used, are the victims, propagators, and rescuers. Victims are described to be immobilized victims that are severely injured or trapped in damaged infrastructures. Meanwhile, propagators are mobile victims who are capable of traversing the damaged area. Lastly, rescuers are mobile individuals who have the complete authority for searching and rescuing especially the immobilized victims. Each of these roles labeled accordingly in the diagram.

At the moment when the immobilized person realizes that rescue is needed, he or she will begin to create the SOS message. When rescuers or propagators start to survey the damaged areas, it is to be assumed that they begin to broadcast for pairing requests. Then the immobilized person, who is severely and may be trapped inside the damaged buildings, will start to listen and scan for the broadcasted pairing requests from propagators that may be passing by as depicted in i of **Figure 1**. Upon capture of a scan packet, the smartphone of the immobilized person will initiate connection with that of the propagator's smartphone and start to communicate. To do so, the

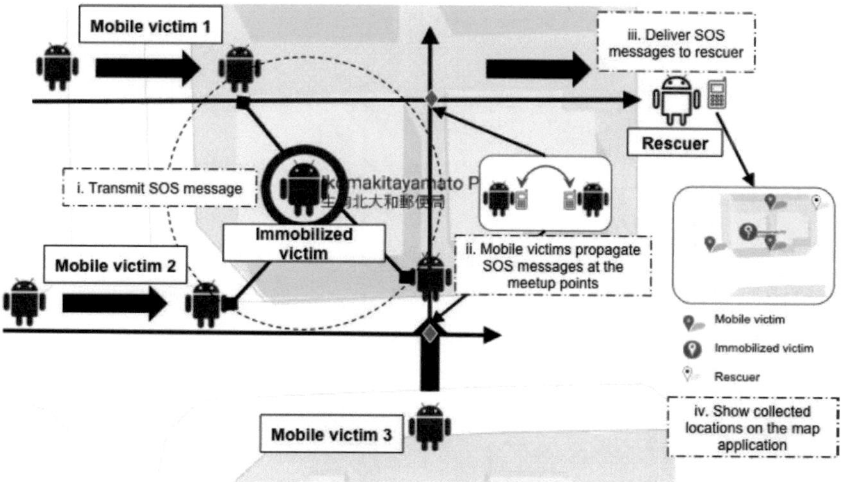

Figure 1. General overview of SOSCast.

immobilized person transmits the created SOS message, which includes the current location based on GPS information. The SOS message composed also states the current condition that the immobilized victim is in and most importantly the victim's name. When the propagator receives the message, an acknowledgement receipt shall be sent back to the victim which states the GPS location information where the propagator has received the SOS message. The propagator's smartphone will then disconnect from the current pairing connection and search for more pairing requests while traversing the vicinity of the damaged area. Such process, when in a repeated cycle, enables the propagator to collect several SOS messages from immobilized victims with successful pairing connections. The propagators are also able to share the received SOS messages with other propagators, such that, each of the propagators will have a consolidated list of SOS messages as in ii of **Figure 1**. After that, it will now be a matter of which propagator is able to deliver the collected SOS messages to rescuers who will estimate the locations of the immobilized victims based on the GPS locations of the SOS messages collected (see iii of **Figure 1**).

Meanwhile, **Figure 2** describes in detail how SOSCast functions on the application level. First, the victim identifies self if immobilized or mobile. Note again that, a victim is considered immobilized if trapped inside a damaged building or infrastructure and is unable to move away from current location. Whereas, a victim is mobile if the individual is capable of traversing the affected area. Now, if the victim is immobilized, the victim will be asked to create the SOS message using the smartphone application with the needed information. Otherwise, the mobile victim would simply enable smartphone to broadcast a pairing request. After the immobilized victim creates the SOS message, the smartphone must be enabled to broadcast a pairing request as well. While the application is used by a mobile victim, the device continues to listen to pairing requests from other victims using the application. If the mobile victim's device found a pairing request, the device will begin to establish connection. Then, if the connection is identified to be with an immobilized victim, the mobile victim's device will wait to receive

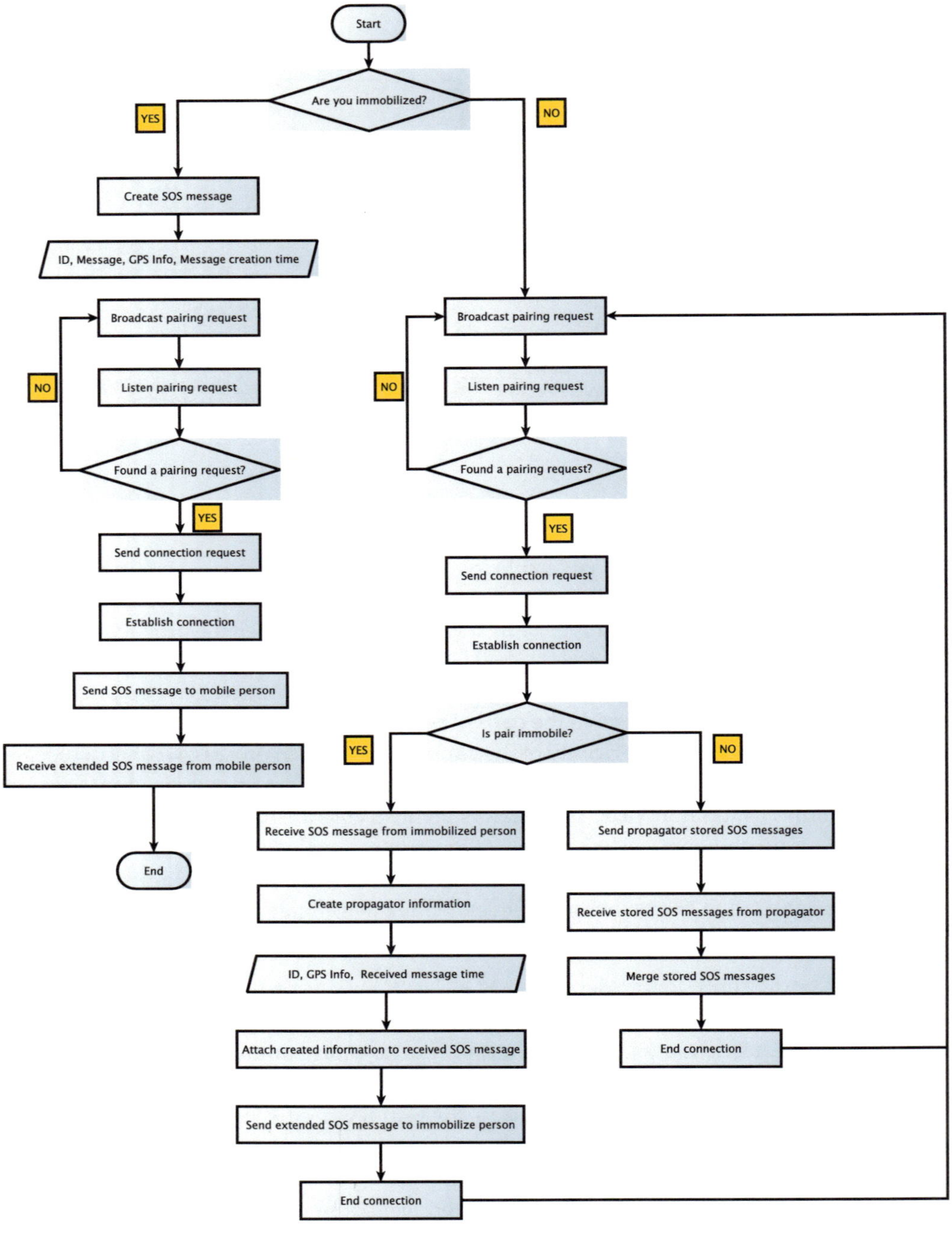

Figure 2. The working process of the SOSCast application.

the SOS message. Simultaneously, the immobilized victim sends the created SOS message to the mobile victim and waits to receive the extended SOS message. At the time when the mobile victim receives the SOS message, the mobile victim will attach personal information to the message, store this on the device, then send back to the immobilized victim the extended message. The purpose of an extended message is for the immobilized victim to confirm that the SOS message has been acknowledged. In that regard, the extended message actually includes the name of the mobile victim and an ID as well as the communicated time and location information. Finally, the mobile victim ends the connection with the immobilized victim and broadcast again a new pairing request searching for other victims.

On the other hand, if the mobile victim happens to find a pairing request from another mobile victim or rescuer, the mobile victim on inquiry mode sends a connection request and establish connection with the other party. Then, instead of receiving an SOS message, the mobile victim sends the other mobile victim or rescuer the currently stored SOS messages from immobilized victims with the extended information. Afterwards, the mobile victim waits to receive the stored messages from the other party to update the database in the device. When the exchange of stored messages is done, both parties end the connection and begin to search for other unaccounted victims.

3.2. Message format

The SOS message formats between immobilized and mobile victims differ in content. The immobilized victim records ID, and remarks on current physical condition. The location information and the time when the message was created are also logged. The immobilized victim's ID may include a nickname or the real name of the person. As for the remarks, it is a message field where the immobilized victim selects from a dropdown menu of current statuses with predefined messages. The location field includes the currently identified GPS information by the device. Lastly, the time when the immobilized victim created the message is also recorded. From this, it is possible to estimate the time when the immobilized victim began to ask for help. Moreover, it is useful to sort and delete duplicated messages based on this time field. **Table 1** lists fields of the message format for immobilized victims with corresponding byte size.

The mobile victim's information includes ID, the current location information, and the time when the mobile victim communicated with the immobilized victim. Similar to immobilized victim's ID, this field may also include a nickname or real name. The location field logs the current GPS information where the mobile victim has established connection with the immobilized victim. Occasionally, the GPS information of immobilized victim may not be accurate or may not be obtained due to indoor location, etc. If so, the information may lead to the wrong operation. However, as the number of obtained messages increases, using the GPS

Item	Description	Size (byte)
Immobilized victim ID (IMEI)	Name to identify the victim	15
Message creation time	Time when the SOS message was composed	4
Immobilized victim location	GPS information of the victim	8
Communication time	Time when a connection was established	4
Propagator location	GPS information of the propagator	8
Immobilized victim information	Current status of the immobilized victim	0

Table 1. Description of the SOS message format.

information of mobile victims improves the precision even if some GPS information of mobile victims are not accurate. Lastly, the time when the mobile victim has received SOS message from the immobilized victim is logged in the time field.

In general, the GPS information for both the immobilized and mobile victims is inaccurate by around 10 m. The SOSCast application is not capable of identifying the accuracy of the obtained GPS information. However, we rely on this information to at least have an idea where the immobilized victim may be at. We discuss this idea further in Section 3.5 to justify how this information proves to be useful even if it is not accurate sometimes.

3.3. Propagation process

When the communication channel is unreliable or unavailable, the smartphones would need to propagate SOS messages by means of store-carry-forward (SCF) routing. In propagating SOS messages, this study mainly implements Bluetooth technology, which is favorable for its simplicity and availability in the majority of smartphones. Bluetooth mainly has two mode types, the inquiry mode and the discoverable mode. In the inquiry mode, a smartphone broadcasts inquiry packets to detect other bluetooth devices. On the other hand, during the discoverable mode, it is listening for broadcast inquiry packets. For the immobilized victims to conserve as much battery as possible, their smartphones are set to discoverable mode. Meanwhile, the propagators could be both in inquiry and discoverable mode, periodically sending inquiry packets during the inquiry mode and constantly listening to broadcast requests during discoverable mode. When the inquiry responses from both immobilized and propagators are received within the period, the propagator generates a list of possible connections. These connections are classified according to the capability of the immobilized person or the propagator to propagate SOS messages as soon as possible.

Table 2 shows six types of connection lists, namely, immobilized persons, propagators, and already connected persons classified into either having a strong or weak Bluetooth Received Signal Strength Indication (RSSI).

Immobilized victims and propagators alike are considered to be already connected persons, as such, persons connected via SOSCast. It is also to be noted that SOSCast in a propagator's smartphone is capable of logging the RSSI of the pairing requests received. With this, the

Priority	Classification	RSSI level
1	Immobilized persons	Strong
2	Propagators	Strong
3	Already connected persons	Strong
4	Immobilized persons	Weak
5	Propagators	Weak
6	Already connected persons	Weak

Table 2. List of SOSCast connections.

propagator can classify the SOSCast connections by the RSSI level. For instance, when the RSSI of the request sent by the immobilized victim's smartphone is higher than the previously set threshold,[1] SOSCast will identify such ID as immobilized persons with strong RSSI. Else, if the identified RSSI is lower than the threshold, then it will be assigned as immobilized persons with weak RSSI. Moreover, propagators and already connected persons are similarly classified with their RSSI relative to the threshold. Propagators with RSSI higher and lower than the threshold are addressed as propagators with strong RSSI and propagators with weak RSSI, respectively. Already connected persons with RSSI higher and lower than the threshold are, namely, already connected persons with strong RSSI and already connected persons with weak RSSI. When such classification list is made, the priority of connection is in the order of strong RSSI from the immobilized person, propagators and already connected persons. Within the classification, the order of addressing which connection will depend on the order the IDs whose SOS message arrive first. Less priority is given to those IDs with weak RSSI beginning from immobilized persons, propagators and already connected persons. Connections to persons with weak RSSI will be postponed because there is a lower probability that a connection can be made compared with those having higher RSSIs.

3.4. Message deletion

Redundant SOS messages are to be deleted properly from the network to avoid misleading the rescuers to continue the search when the immobilized person was already rescued. Removal of such messages must be done immediately for time efficiency during rescue. Illustrated in **Figures 3** and **4** are the deletion processes of unnecessary SOS messages.

In **Figure 3**, assuming that the immobilized person has been rescued and if such person is capable of moving around to help in the rescue, the status will be changed to propagator.

This person will now be propagating other immobilized persons' SOS messages the same way as propagators do. In the status transition, however, the information sent to the consolidated SOS message list must be deleted and should indicate self as "RESCUED" in the message field. The now mobile victim should be able to collect SOS messages from immobilized persons and exchange information with other propagators. Within the collected SOS messages from other propagators, the information relating to self upon the change in status must be removed as in **Figure 4**.

However, some victims who had sent SOS messages and were rescued may need to call rescue again by some reason such as during a secondary disaster. They can also create SOS messages, and these messages are not deleted by the old rescued message. Note that the rescued information is maintained in the SOS message. Therefore, rescued immobilized persons' information is deleted during propagating process.

3.5. Location mapping

Based on the logged SOS messages, the rescuers can easily find the existence of the immobilized victims from the recorded GPS location information. This is so because the SOS messages

[1]This threshold was determined through a preliminary experiment to check whether the connection was stable.

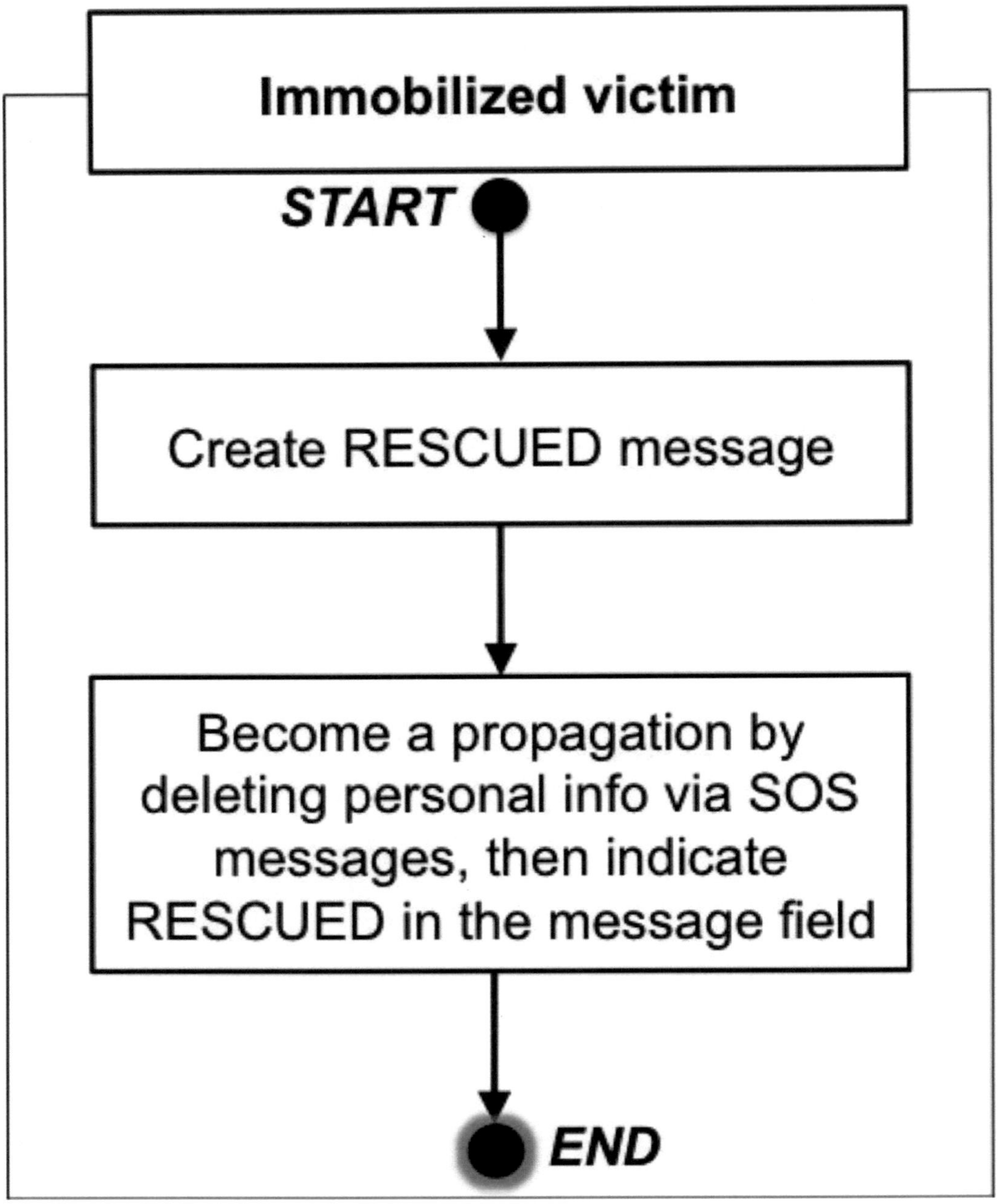

Figure 3. Deletion process in immobilized victim's device.

will be propagating within the network of propagators until the consolidated message list is received by the rescuers. Rescuers and propagators or mobile victims alike carry the same function of the SOSCast in their smartphones. However, only the rescuers has the capability of actually searching for the immobilized victims and the authority of rescuing them. The map

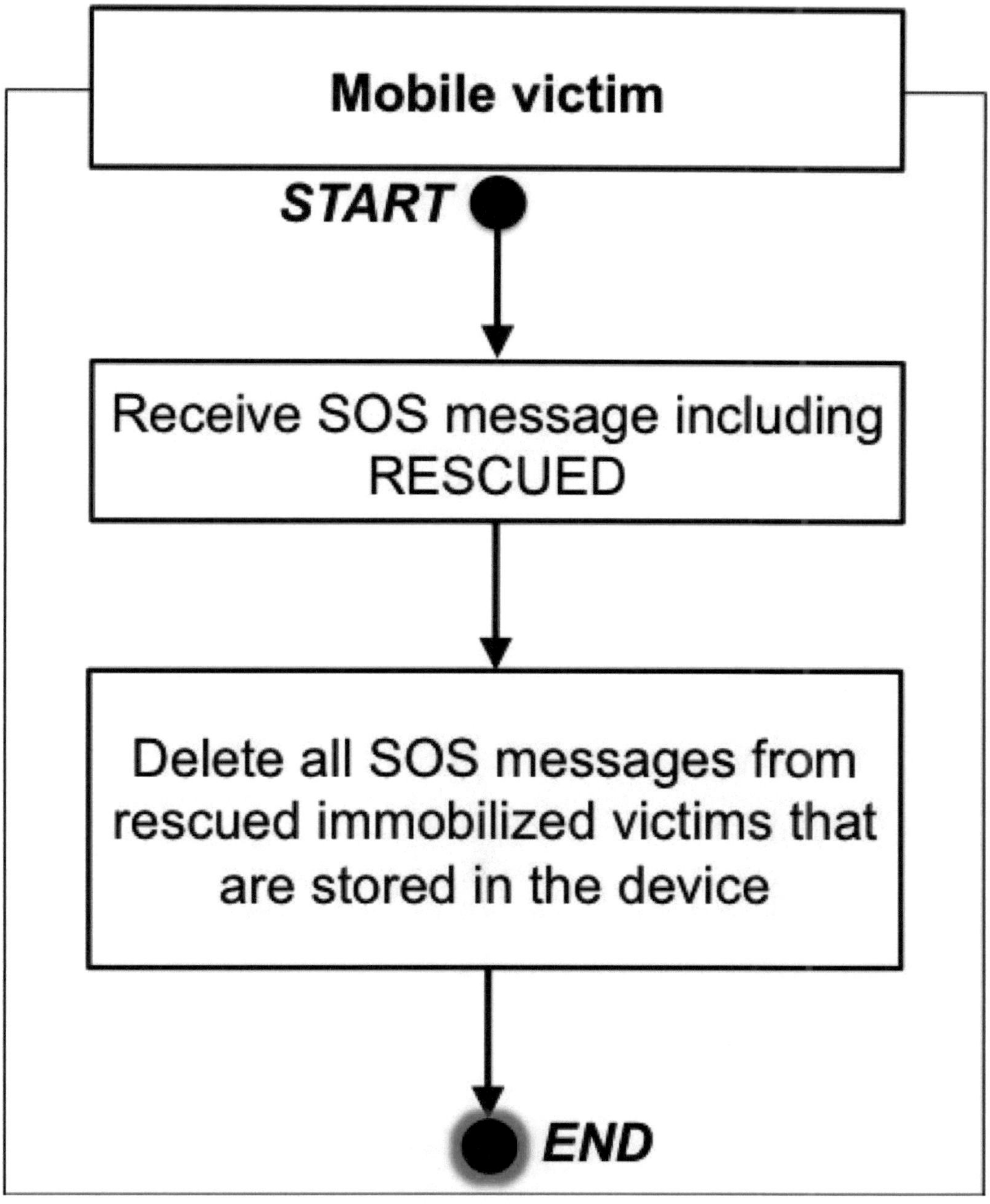

Figure 4. Deletion process in mobile victim or propagator's device.

in SOSCast, meanwhile, helps the rescuers by giving an idea on where the immobilized victim should be searched. Note that SOSCast also collects the GPS location from where the propagator has received the SOS message as an acknowledgement. Without such, it will be difficult to know where to search for the immobilized victim when the given GPS location information of the immobilized victim is unavailable or incorrect. It would be helpful, thus, to have both the acknowledged receipt location from the propagators and the location the immobilized victim provided to make it easy for the rescuers to estimate the search area efficiently.

4. Evaluation

To examine how a smartphone obtain messages in the real environment, we conducted an experiment using Android OS-based smartphones. Section 4.1 describes the experimental environment while Section 4.2 discusses the results.

4.1. Environment

Based on the design of the SOSCast in the general overview, the application was implemented based on Android. To evaluate the potential of the Android application, a preliminary experiment within a residential area nearby the campus was conducted. All persons assuming the roles of the immobilized person and propagator in the experiment have SOSCast installed on their respective device. In this case, the Samsung Galaxy Tab SC-01C was used as a smartphone device by both the immobilized person and propagator. In the scenario as in **Figure 5**,

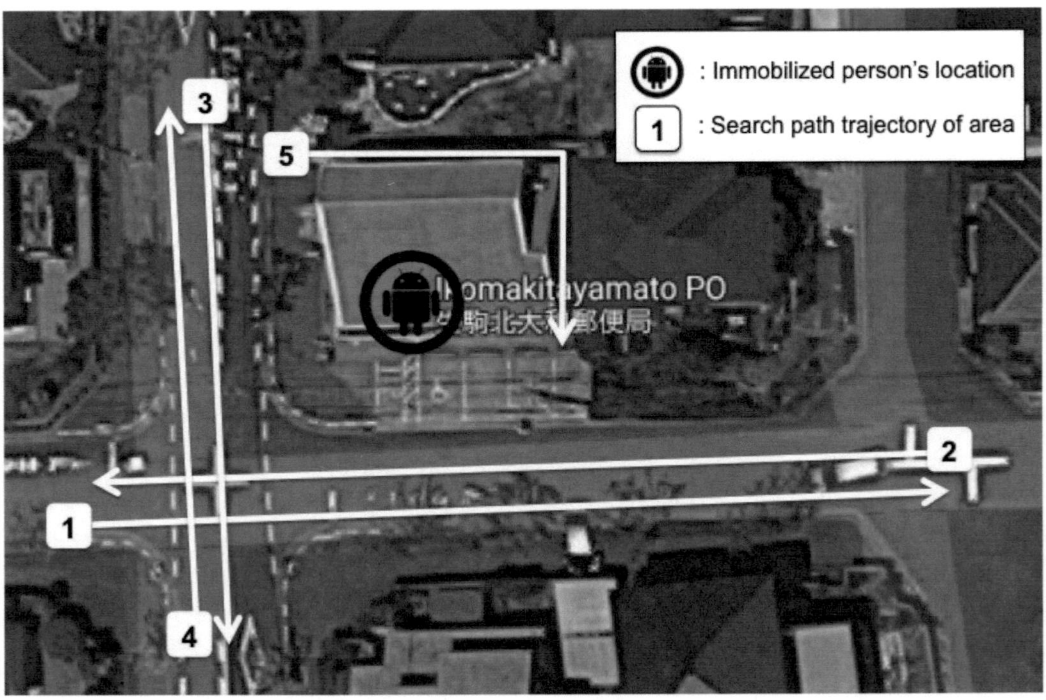

Figure 5. Experimental environment.

an immobilized person is assumed to be trapped inside the building with propagators moving around the building. Considering that the building is damaged and inaccessible, propagators walk along the surrounding street. Traversing the area is done five time in all five directions indicated. When a rescuer receives the collected SOS messages from the propagators, the logged GPS location information from the SOS messages both indicated by the immobilized persons and that of the location acknowledged by the propagator of having received the message will be displayed on the map.

4.2. Results

Figure 6 is a screenshot of the display on the rescuer's smartphone which shows the locations of both the immobilized person and propagators. The results of the experiment have shown us that the propagators were able to receive the messages within the 10 m parameter in average on the directions indicated in 1, 2, and 5. In the case of directions 3 and 4, the distances of the locations where the SOS messages were received are higher than average. Even so, the availability of the distance, based from those indicated by the propagator from where the SOS messages were acknowledged to have been received, could indicate the presence of an immobilized person within the vicinity. Again, this could aid the rescuers in estimating the search area even with incorrect or unavailable GPS location information from immobilized person.

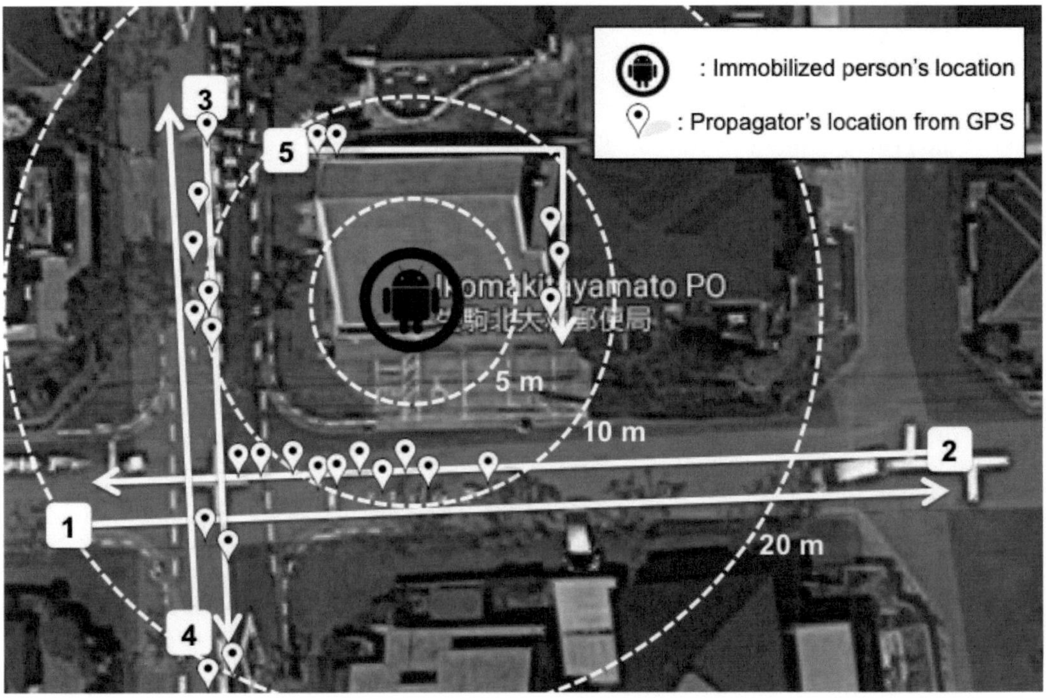

Figure 6. Experimental result.

5. Conclusion and future work

In this chapter, we showed the implementation and design of sending SOS messages by both immobile and mobile victims of a very damaging disaster, namely the SOSCast application. By estimating the locations, we have shown that SOSCast has the potential to locate victims specifically targeting immobilized persons trapped under fallen debris or infrastructures based on this information. In SOSCast, the victims collect and propagate these messages which rescuers can use to estimate locations of the immobilized persons by using the smartphone. The more GPS information received from collected SOS messages, the more it is possible to estimate the victim's location. In a disaster area where conventional communication services are impaired, SOSCast can potentially aid the work of rescuers as it can enable direct communication between smartphones. As future work, we will consider to utilize other wireless communication like Wi-Fi Direct, in addition to Bluetooth. Also, with the development of Bluetooth Low Energy technology, this could dramatically reduce power consumption during pairing request broadcasts between immobilized persons and propagators.

Acknowledgements

This work was supported by KAKENHI (26730050 and 17K00124).

Author details

Jane Louie Fresco Zamora[1]*, Noriyuki Suzuki[2] and Shigeru Kashihara[1]

*Address all correspondence to: louie-z@is.aist-nara.ac.jp

1 Nara Institute of Science and Technology, Ikoma, Japan

2 Internet Initiative Japan, Inc, Tokyo, Japan

References

[1] Ministry of Internal Affairs and Communications (Japan). WHITE PAPER Information and Communications in Japan 2011. URL http://www.soumu.go.jp/johotsusintokei/whitepaper/eng/WP2011/2011-index.html

[2] Jang H-C, Lien Y-N, Tsai T-C. Rescue information system for earthquake disasters based on manet emergency communication platform. In Proceedings of the 2009 International Conference on Wireless Communications and Mobile Computing: Connecting the World Wirelessly, IWCMC '09; New York, NY, USA. ACM; 2009. pp. 623-627. ISBN 978-1-60558-569-7. DOI: 10.1145/1582379.1582514

[3] Wu X, et al. Emergency message dissemination system for smartphones during natural disasters. In Proceedings of ITS Telecommunications (ITST). 2011; pp. 258-263

[4] Teranishi Y, Shimojo S. Monac: Sns message dissemination over smartphone-based dtn and cloud. In: 2011 IEEE International Conference on Peer-to-Peer Computing (P2P); 2011. pp. 158-159. DOI: 10.1109/P2P.2011.6038726

[5] Sakurai M, Watson RT, Abraham C, Kokuryo J. Sustaining life during the early stages of disaster relief with a frugal information system: Learning from the great east Japan earthquake. IEEE Communications Magazine. January 2014;**52**(1):176-185. ISSN 0163-6804. DOI: 10.1109/MCOM.2014.6710081

[6] Sakano T, Fadlullah ZM, Thuan Ngo, Nishiyama H, Nakazawa M, Adachi F, Kato N, Takahara A, Kumagai T, Kasahara H, Kurihara S. Disaster-resilient networking: A new vision based on movable and deployable resource units. IEEE Network. July 2013;**27**(4):40-46. ISSN 0890-8044. DOI: 10.1109/MNET.2013.6574664

[7] Nishiyama H, Ito M, Kato N. Relay-by-smartphone: Realizing multihop device-to-device communications. IEEE Communications Magazine. April 2014;**52**(4):56-65. ISSN 0163-6804. DOI: 10.1109/MCOM.2014.6807947

[8] Chipara O, Griswold WG, Plymoth AN, Fang Liu RH, Johansson P, Rao R. Chan T, Buono C. Wiisard: A measurement study of network properties and protocol reliability during an emergency response. In: Proceedings of the 10th International Conference on Mobile Systems, Applications, and Services, MobiSys '12; New York, NY, USA: ACM; 2012. pp. 407-420. ISBN 978-1-4503-1301-8. DOI: 10.1145/2307636.2307674

[9] Chenji H, Zhang W, Stoleru R, Arnett C. Distressnet: A disaster response system providing constant availability cloud-like services. Ad Hoc Networks. 2013;**11**(8):2440-2460. ISSN 1570-8705. DOI: http://dx.doi.org/10.1016/j.adhoc.2013.06.008. URL http://www.sciencedirect.com/science/article/pii/S1570870513001315

Chapter 4

Smartphone as a Portable Detector, Analytical Device, or Instrument Interface

Diana Bueno Hernández, Jean Louis Marty and
Roberto Muñoz Guerrero

Additional information is available at the end of the chapter

http://dx.doi.org/10.5772/intechopen.69678

Abstract

The Encyclopedia Britannia defines a smartphone as a mobile telephone with a display screen, at the same time serves as a pocket watch, calendar, addresses book and calculator and uses its own operating system (OS). A smartphone is considered as a mobile telephone integrated to a handheld computer. As the market matured, solid-state computer memory and integrated circuits became less expensive over the following decade, smartphone became more computer-like, and more more-advanced services, and became ubiquitous with the introduction of mobile phone networks. The communication takes place for sending and receiving photographs, music, video clips, e-mails and more. The growing capabilities of handheld devices and transmission protocols have enabled a growing number of applications. The integration of camera, access Wi-Fi, payments, augmented reality or the global position system (GPS) are features that have been used for science because the users of smartphone have risen all over the world. This chapter deals with the importance of one of the most common communication channels, the smartphone and how it impregnates in the science. The technological characteristics of this device make it a useful tool in social sciences, medicine, chemistry, detections of contaminants, pesticides, drugs or others, like so detection of signals or image.

Keywords: smartphone, detection, chemistry, optical, medicine, mobile applications, instrumental interface

1. Introduction

Smartphones are similar to notebook computer with its own operating system, processor, internal memory, and high-quality camera lenses [1]. The smartphones are more accessible and cheaper than portable analytical devices. According to eMarketer, the number of smartphone users worldwide surpassed 2 billion in 2016 and it represents more than a quarter of the global population. For 2018, the mobile users will grow to more than 2.56 billion people or a third of the world's population. However, the latest annual Mobility Report from Ericsson indicated that nowadays, there are 84 million new mobile subscriptions, reaching a total of 3.9 billion of smartphone subscriptions. Mobile subscriptions are growing at around 3 percent year-on-year globally. And it have been estimated that there will be a 6.8 billion of smartphone subscriptions for 2022. This growth has a big impact, and it highlights the opportunity to create apps and other services to meet our necessities in a practical way. The smartphone ownership rates in emerging and developing nations are rising, 21% in 2013 and 35% in 2015. **Figure 1** shows a distribution of the smartphone users by countries [2].

The continuous improvement of smartphone electronics, the development of new app and the increase of users have stimulated research in the use of smartphone. Smartphone technology now includes a range of detection capabilities, thanks to the built-in camera, such as colorimetric detection [3], optical methods that are employed in an easy way and to share the information on real time, as well as quantification [4], monitoring [5], or mobile applications [6–8] developed to solve a problem such as freezing of gait in Parkinson's disease [9], glucose monitor [10], and to detect enzymes [11, 12]. Also, it is used to developed devices, a

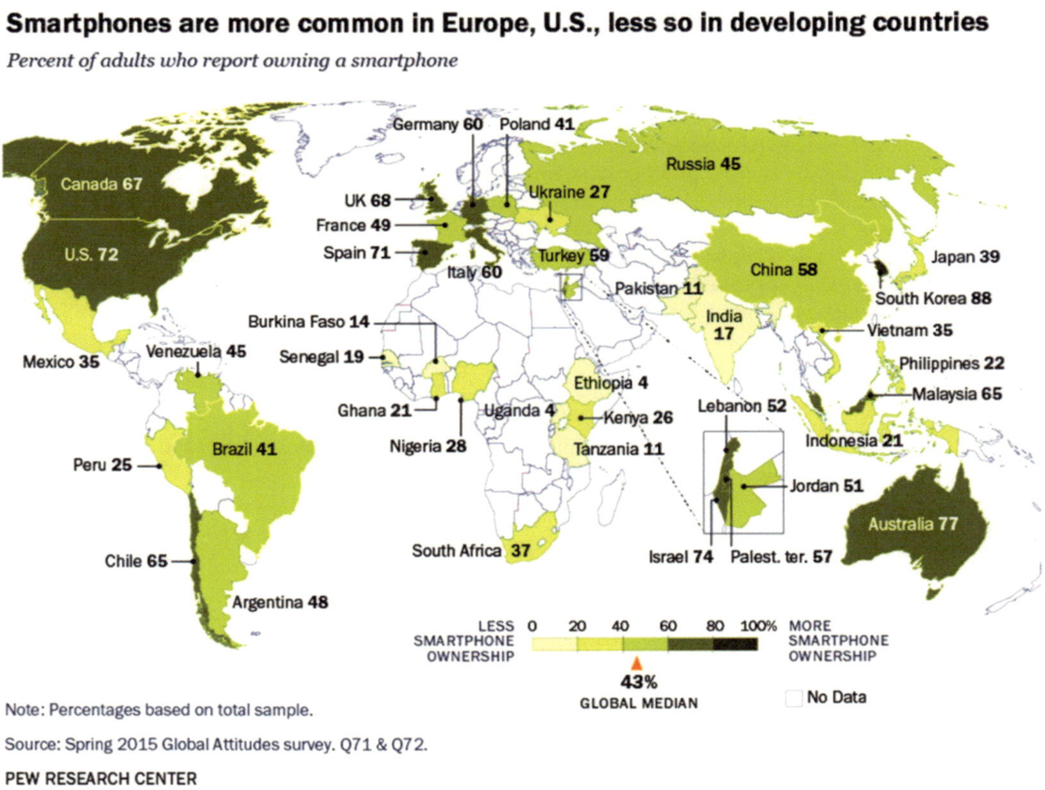

Figure 1. Smartphone's users in worldwide.

spectrophotometer [13, 14] and an electrocardiograph [15], and to create sensors or biosensors [1]. Some applications in healthcare, are: melanoma detection [16], cancer prevention [17], emergency signal detection [18], optical imaging techniques for diagnostics [19, 20], or detection [21]. Besides, in the environment, the smartphone has been used to analyze the quality of the water [22] to detect its salinity [23] or to detect mycotoxins [24, 25]. A complete revision of the applications in different area, devices developed, and limit of detection (LOD) of difference samples will be included in this review.

In this review has been illustrated the smartphone as a research tool to the detection and analysis, because these can be employed in interdisciplinary areas. One the most representative characteristics of the smartphone is that it allows the portability because of its size and its cost compared with commercials instruments. The smartphone sensors have the capacity to effectively serve as portable biofeedback devices for a diverse range of applications in science. Its processor allows you to collect, analyze, and process images or signals in an embedded way.

2. Applications of the smartphone

There have been many recent publications on the use of smartphone as portable detector, bioanalytical devices and instruments interfaces, among others. The use of smartphone had created opportunities for diagnostic, prognostic, detection, quantification, monitoring, control or make mobile applications, because it could be used to run routine test, does not need trained personal, its portability and is considered as a low cost device. Moreover, to have a complete overview of the real advantages, applications or characteristics of the smartphone, it is important to point out the problems with the sampling of real samples (biological, food, environmental), and these need conventional devices to compare the results.

Sometimes, a smartphone is coupled with a device that contains the components of an instrument in a simplified format. For analytical applications, the smartphone is used to control the experimental device and display the results on a dashboard in a tablet, television, computer or other screen and to communicate via USB port, Bluetooth or Wi-Fi between the smartphone and the analytical device. Some examples are described as follows.

2.1. Medicine

Smartphone offers potential for medical diagnosis [26, 27] and treatment of pathologies as a low cost system. There is an increasing interest to detect analytes of clinical interest as employing the mobile phone camera for DNA detection [28], DNA amplification uses the convective polymerase chain reaction technique, and the detection is carried out with the variation in the fluorescence. The fluorescence increment used the brightness of the image before and after the DNA amplification. If there is a difference between before and after the DNA amplification, the test is positive. This process can be used for screening hepatitis B virus plasmid samples.

In 2014, Guan et al. introduced a barcode design into a paper-based blood typing device by integrating with smartphone, and this device involved the use of hydrophilic bar channels treated with anti-A, B and D antibodies. These channels were then used to perform blood typing assays by introducing a blood sample. Blood type can be visually identified from eluting lengths in bar channels. A smartphone application was designed to read the bar channels, analogous to scanning a barcode, interpret this information, and then report the results to users [29].

The development of a microscope attached to smartphone was reported by Breslauer et al., and the authors demonstrated the applicability of this device for clinical diagnostics of *P. Falciparum* and *M. tuberculosis*, providing an important tool for disease diagnosis and screening, particularly in the developing world and rural areas where laboratory facilities are scarce, but mobile-phone infrastructure is extensive [30].

Surface plasmon resonance (SPR) detection system based on a smartphone was proposed by Preechaburana et al. where the authors demonstrated that the resolution of the device employing the smartphone is comparable with conventional analytical SPR. The assays were made for the detection of β2 microglobulin, biomarker for cancer, and were achieved a limit of detection of 0.1 mg/mL in urine [31]. More works about the use of the smartphone are presented in **Table 1**, where the samples detected and brief descriptions of the work are mentioned.

2.2. Chemistry

Nielsen reports that between 2009 and 2011 smartphone ownership for 13–17 year olds and 18–24 year olds went from 16 to 40% and from 23 to 53%, respectively, an increase of 100% approximately. Smartphone serves as powerful educational tool on a mobile platform, which encourages learning. The mobile applications, or "apps," have a wide range of functionalities and cover many disciplines. Collaboration through the interconnection of multiple chemistry apps has demonstrated that chemo-informatics is a tool to increase work efficiency, which can be utilized to raise the chemistry learning experience to a new level, and Chemspider app is a powerful handheld chemical search engine. This chapter [46] discusses apps that are available on smartphone, as these are more prevalent, affordable, and portable than comparable tablets or laptops.

A novel approach for an inexpensive and disposable colorimetric paper sensor array for the detection and discrimination of five explosives (triacetone triperoxide, hexamethylene triperoxide diamine, 4-amino-2-nitrophenol, nitrobenzene and picric acid) was presented by Salles et al. [47]. The colorimetric sensor was designed as a disposable paper array fabricated with potassium iodide, creatinine and aniline, which produces a chemical reaction, a specific color pattern for each explosive. The analytes were identified and classified for each explosive by the changes in the color patterns, which were extracted using a smartphone camera installed in a closed chamber to avoid the illumination interactions. A semiquantitative analysis was performed, and it was possible to use as low as 0.2 mg of explosives. Others detections with the smartphone are presented in **Table 2**.

Samples	Detection/quantification	Short description	References
Vitamin D	Measure physiological levels of 25-hydroxyvitamin D with accuracy better than 15 nM and a precision of 10 nM	The system consists of a smartphone accessory, an app and a test strip that allows the colorimetric detection of 25-hydroxyvitamin D using a novel gold nanoparticle-based immunoassay	[4]
Freezing of gait (FOG) is a motor symptom in patients with Parkinson's disease (PD)	98 FOG events were recognized, and seven FOG events were missed by the application. Sensitivity and specificity were 70.1 and 84.1%, for the Moore-Bächlin Algorithm, rising to 87.57 and 94.97%, for the second algorithm employed	In order to verify the acceptance of a smartphone-based architecture and its reliability at detecting FOG in real-time, 20 patients were studied. It consisted to make a video-recorded Timed Up and Go using the smartphone; the video was synchronized with the accelerometer to assess the reliability of the FOG detection system as compared to the clinicians. The algorithms employed were the Freezing and Energy Index (Moore-Bächlin Algorithm)	[9]
Salivary glucose	Detection range of 9–1350 mg/dL glucose at a response time of 45 s and LOD of 22.2 mg/dL	The assay consisted to immobilized glucose oxidase enzyme on filter paper strip (specific activity 1.4 U/strip); the enzyme reacts with synthetic glucose samples in presence of co-immobilized color pH indicator. Then, the changes in the filter paper based on concentration of glucose are detected. Once the biosensor was standardized, the synthetic glucose sample was replaced with human saliva	[32]
Salivary cortisol	LOD of 0.3 ng/mL. It provides quantitative analysis in the range of 0.3–60 ng/mL	The biosensor is based on a direct competitive immunoassay with peroxidase–cortisol conjugate by adding the substrate luminol/enhancer/hydrogen peroxide, and it produces a chemiluminescence reaction. The smartphone was used to detect the light generates by the reaction through the camera as image and the data handling via an application	[33]
Blood hematocrit	0.1% of hematocrit with a sensitivity of 0.53 GSV (a.u.)/hematocrit	Using an integrated camera in the smartphone, pictures of human blood in the microchannel were taken and analyzed by a mobile application. The characterization of the depth of the microfluidic channel demonstrated that a shallower depth of the microchannel enhanced the sensitivity of the hematocrit determination	[34]
Albumin in urine	5–10 μg/mL^{-1} (which is more than three times lower than the clinically accepted normal range)	The test and control tubes are excited by a battery powered laser diode; its fluorescent emission is collected perpendicular to the excitation by a smartphone camera, through external plastic lens inserted in the camera lens. The images are digitally processed within one second through an Android application, with the purpose to quantify the albumin concentration in the urine samples	[35]

Samples	Detection/quantification	Short description	References
Thyroid stimulating hormone (TSH)	0.31 mIU/L	Employing the methodology of the optimized Rayleigh/Mie scatter detection with the optical characteristics of a nitrocellulose membrane and gold nanoparticles for quantifying TSH levels. Using A miniature spectrometer, light-emitting diodes (LED) as light source and optical fibers on a rotating benchtop apparatus, the light intensity from different angles of incident light and angles of detection were measured. A bracket was designed to support the cell-phone and the embedded flash as the light source, through a collimating lens to illuminate the assays, and quantified the concentration of TSH in an iOS application, and it was verified using a code made in MATLAB	[36]
Cholesterol in blood	Cholesterol levels within 1.8% accuracy in the relevant physiological range (140 mg dl^{-1} to 400 mg dl^{-1})	A smartcard, smartphone Cholesterol Application for Rapid Diagnostics system. The system can quantify cholesterol levels from colorimetric changes due to cholesterol reacting enzymatically on a dry reagent test strip. The smartphone acquires the image of the test strip and an app that analyzes parameters such as hue, saturation and luminosity of the test area quantifies the cholesterol levels and displays the value on the screen	[37]
Glucose and urea	Glucose had a concentration ranges 30–515 mg/dl and 2–190 mg/dl for blood urea nitrogen (BUN)	Smartphone equipped with a color analysis application was combined with Vitros® glucose and urea colorimetric assays. Color images of assay slides at various concentrations of glucose or urea were collected and quantitated in three different spectral ranges (red/green/blue or RGB). When the diffuse reflectance data were converted into absorbance, it was possible to quantitate glucose or BUN	[38]
Kaposi's sarcoma	It can detect DNA sequences from KSHV down to 1 nM	Kaposi's sarcoma (KS) is an infectious cancer occurring in immune-compromised patients, caused by Kaposi's sarcoma associated herpesvirus (KSHV). In this work, a smartphone accessory capable of detecting KSHV nucleic acids was developed. The accessory reads out microfluidic chips filled with a colorimetric nanoparticle assay targeted at KSHV	[39]
DNA molecules	Longer DNA samples imaged over a field-of-view of 2 mm	The images and the length quantification of single-molecule DNA strands using a fluorescence microscope installed on a mobile phone. An optomechanical bracket with lens, thin-film interference filters, laser-diode and a mobile phone application were designed to measure the lengths of DNA molecules labeled and stretched using disposable chip	[40]

http://dx.doi.org/10.5772/intechopen.69678

Samples	Detection/quantification	Short description	References
DNA/RNA analysis	Replication of Staphylococcus aureus and λ-phage DNA targets in less than 20 min, in-flight	This approach exploits the ability to isothermally perform the polymerase chain reaction (PCR) with a single heater, enabling the system to be operated using standard 5 V USB sources that power mobile devices. Time-resolved fluorescence detection and quantification is achieved using a smartphone camera and integrated image analysis app	[41]
Glucose	~3.5μm	Khan et al. described a label-free, optical sensor capable to detect the changes in photoluminescence (PL) of a thin polymer film employing the glucose as the target molecule, the radiance changes of the quantum dots in PL of the UV excitation of the enzymatic reaction, allows quantifying the level of glucose	[42]
Protein	Bovine serum albumin (BSA) of 1.78μg/ml Thrombin of 2.97ng/ml	A smartphone-controlled biosensor system is developed with electrochemical impedance spectroscopy (EIS) to detect proteins for POC testing. Printed carbon electrodes and interdigital gold electrodes were modified with biocomponents as biosensors for quantification of the protein binding and enzyme activities by special antibodies and peptides immobilized on the electrodes	[43]
pH in sweat and saliva	Unmentioned	A smartphone camera and colorimetric detection of pH in sweat and saliva. Sweat pH can be correlated to sodium concentration and avoid the risk of muscle cramps. Salivary pH below a critical threshold is correlated with enamel decalcification, an acidic breakdown of calcium in the teeth	[44]
Escherichia coli and gonorrhea in human urine	10 CFU/mL for both	Anti-*Escherichia coli* or anti-*Neisseria gonorrhoeae* antibodies were conjugated to submicron particles. The bacteria-spiked urine samples were introduced to the inlet of paper microfluidic channel, which flowed through the channel by capillary force. Urobilin, the component responsible for the yellow appearance of urine and green fluorescence emission, was filtered by a microfluidic paper analytical device (μPAD). The extent of immunoagglutination was quantified by angle-specific Mie scatter under ambient lighting conditions, utilizing a smartphone camera as a detector	[45]

Table 1. Uses of the smartphone in medicine.

Samples	Detection/quantification	Short description	References
Lactate in oral fluid and sweat	0.5 mmol L^{-1} in oral fluid and 0.1 mmol L^{-1} in sweat	To develop a portable lactate chemiluminescent biosensor, based on the coupling of the enzymatic oxidation of lactate catalyzed by L-lactate oxidase with the luminol/H2O2/HRP CL system, using disposable analytical cartridges to allow measurement of the light produced by the enzyme reaction and a smartphone camera to detect the light	[48]
Liquids	Unmentioned	A handheld automated microfluidic liquid handling system is controlled by a smartphone, which is enabled by combining elastomeric on-chip valves and a compact pneumatic system, and it can automatically perform all the liquid handling steps of a bead-based HIV1 p24 sandwich immunoassay on a multi-layer PDMS chip without any human intervention	[49]
Cocaine	LOD of 0.25 mg/mL	An assay based on the gold nanoparticle conjugate (AuNPs) difference in affinity for single-stranded DNA (nonbinding) and double stranded DNA (target bound). The AuNPs and the aptamer were incubated prior to target addition to passivate the AuNPs surface. The adsorbed aptamer was able to bind the target to avoid nonspecific interactions. To facilitate the assay analysis, an android application for automatic colorimetric characterization was developed	[50]
H_2O_2 Glucose in phosphate buffer Artificial urine	1.75 μM 0.017mM 0.030mM	A handheld paper-based bipolar electrode-electro-chemiluminescence (P-BPE-ECL) system with a rechargeable battery as power supply and smartphone for readout of ECL signal is employed. In the case of the electro-chemiluminescence reaction, the carbon ink-based bipolar electrode and driving electrodes are screen-printed and the wax-screen-printing is employed to fabricate the microfluidic channels. The luminol/H2O2-based ECL reaction is applied to quantify the P-BPE-ECL system	[51]
Heavy metals	Cu (II)=0.29 ppm Ni (II)=0.33 ppm Cd (II)=0.19 ppm Cr (VI)=0.35 ppm	A 3D paper-based microfluidic device for colorimetric determination of selected heavy metals in water samples. The process is as follows: the samples are immersed into the paper chip and the sample streams reaching into the detection zones. Finally, the activation solutions are dropped to get metal chromogenic reactions that are captured by a camera cell phone and analyzed in a personal computer employing image processing and analysis software	[52]
Salmonella spp. *Escherichia coli* O157	10^5 CFU (colony forming units) mL^{-1}	Silica nanoparticles (SiNPs) were doped with FITC and Ru, conjugated to the respective antibodies and used in a conventional lateral flow immunoassay (LFIA). Fluorescence was recorded by inserting the nitrocellulose strip into a smartphone-based fluorimeter consisting of a LED, a fluorescence filter set and a lens attached to the integrated camera. The images were analyzed by exploiting the quick image processing application of the cell phone and enable the detection of pathogens within few minutes	[53]

http://dx.doi.org/10.5772/intechopen.69678

Samples	Detection/quantification	Short description	References
Salmonella	10^2 CFU mL^{-1} with a linear range up to 10^5 CFU mL^{-1}	Each microfluidic channel was preloaded with anti-Salmonella Typhimurium and anti-Escherichia coli conjugated. The paper microfluidic device was submerged into the Salmonella solutions led to the antibody-conjugated particles to immunoagglutinate. The immunoagglutination was quantified by evaluating Mie scattering from the digital images captured. A smartphone application was designed and programmed to allow the user to position the smartphone at specific angle and distance from the microfluidic device. Besides, an image processing algorithm was implemented to quantify the bacterial concentration	[54]
Amines	Less than 1ppm	Bueno et al. present the use of solvent cast cellulose acetate membranes to immobilize dyes and to employ the membranes as a plastic device to identify between different types of amines (triethylamine, isobutyl amine, isopentylamine). The device consisted of an array of membranes with five pH indicators (alizarin, bromophenol blue, chlorophenol red, methyl red and thymol blue). A smartphone was used to analyze the data, to capture the images and to extract the red, green and blue (RGB) components from the image to generate a unique color pattern	[55]

Table 2. Applications of the smartphone in chemistry area.

2.3. Food

A smartphone-utilized biosensor was developed for detecting microbial spoilage (Escherichia coli) on ground beef, without using antibodies, microbeads or any other reagents, toward a preliminary screening tool for microbial contamination on meat products and potentially toward wound infection. Near infrared LED was used to irradiate perpendicular to the surface of ground beef, and the scatter signals at various angles were evaluated utilizing the gyro sensor and the digital camera of a smartphone. The fluorescence microscopy experiments revealed that the antigens and cell fragments from E. coli bonded preferably to the fat particles within meat, and the size and morphologies of such aggregates varied by the E. coli concentration, concluded by Liang et al. [56].

Yu et al. designed a disposable lateral flow-through strip for smartphone to fast one-step quantitatively detect alkaline phosphatase (ALP) activity in raw milk. The strip comprises two functional components, a conjugation pad loaded with phosphotyrosine-coated gold nanoparticles and a testing line coated with anti-phosphotyrosine antibody. The dephosphorylation activity of ALP at the testing zone can be quantitatively assayed by monitoring the accumulated gold nanoparticles and induced color changes by smartphone camera, thus providing a highly convenient portable detection method, demonstrating the potential of smartphone for pathogen detection. Other application of the biohazard-free lateral flow-through testing strip is for the fabrication of rapid, sensitive and inexpensive enzyme or immunosensors for food contamination, food quality inspection or clinic [57]. Allowing the diagnosis or analysis, some elements detected in food are represented in **Table 3**.

Samples	Detection/quantification	Short description	References
L-glutamate dehydrogenase in wine and instant soups	0.5–5.0 mmol L^{-1} for image processing (linear range) The LOD was 0.05 mmol L^{-1} and 0.028 mmol L^{-1} by naked eye and image processing, respectively	A chromatography paper for the analysis of selected food compounds was developed by Monosik et al. The biochemical colorimetric assay utilizes enzymes from the dehydrogenase family coupled with diaphorase in the presence of a tetrazolium dye, MTT and NAD+. The product of the colorimetric reaction developed on the surface of the paper is observed by the naked eye and was captured by smartphone camera to get the data for a quantitative analysis	[58]
Anti-recombinant bovine somatotropin (rbST) in milk	An 80 % true-positive rate and 95 % true-negative rate were achieved	The rbST biomarker present in milk was captured by rbST covalently coupled to paramagnetic microspheres and labeled by quantum dot (QD)-coupled detection antibodies. The emitted fluorescence light from the QDs was captured using the cell phone camera. The fluorescence and dark-field microimages were analyzed using an Android application developed	[59]
Furfural in pale lager beers	12 μgL^{-1}	A disposable color changing polymeric films, the films are prepared by radical polymerization of 4-vinylaniline. The sensitive indicator monomer is the furfural-, the comonomer is 2-hydroxymethyl methacrylate and the cross-linker is the ethylene dimethyl methacrylate (EDMA). As sensing mechanism used the Stenhouse reaction, the aniline and furfural react in acidic media, generating a deep red cyanine product. The colorimetric response has been monitored using either a portable fiber-optic spectrophotometer or the built-in camera of a smartphone	[60]
Tetracycline (TC) in bovine milk	Concentration range of 0.5–10 $\mu g\ mL^{-1}$. LOD of 0.5 $\mu g\ mL^{-1}$ and limit of quantitation of 1.5 $\mu g\ mL^{-1}$	An application named ColorConc was developed for the iPhone that utilizes an image matching algorithm to determine the TC concentration in a solution. The values of red, green, blue, hue, saturation, brightness were measured from each picture. The TC solution extracted from milk samples using solid phase extraction (SPE) was captured and the concentration was predicted by comparing color values with those collected in a database	[61]

Table 3. Smartphone in the detection of food contaminants or food compounds.

2.4. Environment

The contaminated water is a worldwide problem; for that reason, in 2013, Andrade et al. [62] proposed a digital image processing-based flow-batch analyzer for aluminum (III) and chromium (VI) determinations in natural water, employing a webcam with a charge coupled device (CCD) sensor and red, green, and blue (RGB) data. The method for determining aluminum is based on an Al (III) ion, and the reaction produces a yellow-colored complex in an acetate buffer. The determination of chromium is based on a Cr (VI) ion, which produces a violet-colored complex. The digital images were processing, and the RGB data were employed to build the analytical curves. The working ranges were from 10 to 600 μg/L for Al (III) and 10 to 300 μg/L for Cr (VI), and their limits of detection were 3.97 and 2.65 μg/L, respectively, for Al (III) and Cr (VI).

Gopinath et al. have mentioned that the ubiquitous nature of bacteria enables them to survive in a wide variety of environments and provides an overview of the bacterial detection systems that ranges from microscopic observation to smartphone-based detection. It work described that the first application using a smartphone to the detection and to visualize a single bacterium or virus was demonstrated by Zhu et al. [19]. The system was applied to E. coli as a proof-of-concept. Anti-E. coli antibodies were immobilized on the interior surface of a capillary tube [63]. **Table 4** describes some applications of the smartphone in the detection of elements in the environment.

Samples	Detection/ quantification	Short description	References
Catechols from a water sample of a river	Unmentioned	A smartphone-based colorimetric reader was coupled with a remote server for rapid on site analysis of catechols. A colorimetric sensor array composed of pH indicators, and phenylboronic acid was configured. The method identified the catechols with 100% accuracy and predicts the concentrations to within 0.706–2.240 standard deviation	[64]
Mercury contamination in water	3.5 ppb	An optomechanical device integrated to the camera module of a smartphone to quantify mercury concentration using a plasmonic gold nanoparticle (AuNP) and aptamer-based colorimetric transmission assay. It was possible to quantify mercury (II) ion concentration in water samples by using a two-color ratiometric method employing LEDs. Using this smart-phone-based detection platform, we generated a mercury contamination map by measuring water samples at over 50 locations in California (USA), taken from city tap water sources, rivers, lakes, and beaches	[65]
Calcium in water of the city net, mineral bottled, and natural-river	$0.07 mgL^{-1}$	The studies were carried out using the chromogenic model formed by the reaction between Ca (II) ions and glyoxalbis (2-hydroxyanil). It produced orange–red-colored solutions in alkaline media. The colored complex was applied and validated using intensity of colors with factorial analysis, Fourier transform, principal component analysis and the digital image colorimetric method for the determination of Ca (II) ions	[66]
Bacteria in field water	10 bacterial cells per milliliter	The design, fabrication and testing of a low-cost, miniaturized and sensitive bacteria sensor based on electrical impedance spectroscopy method using a smartphone	[67]
Pesticide thiram	0.1 μM	Copper ions decorated NaYF4:Yb/Tm up conversion nanoparticles were fixed onto filter paper for the assay, and the blue luminescence was quenched by the addition of thiram. The differences of blue channel intensities were monitored by the smartphone camera and calculated to quantify amounts of thiram through a self-written Android program installed on the smartphone	[68]

Samples	Detection/ quantification	Short description	References
Trinitrotoluene in soil	50 mgL^{-1}	The built-in digital camera of a smartphone was used to capture the results from a rapid quantitative colorimetric test for trinitrotoluene (TNT) in soil. The colored product from the selective test for TNT was quantified using the relationships between the red, green, blue (RGB) values and the concentrations of colorimetric product	[69]

Table 4. Smartphone in environment detection.

2.5. Biosensors or devices

Smartphone has been widely integrated with sensors, such as test strips, sensor chips and handheld detectors. The biosensors or devices based on smartphone can mainly be classified into biosensors using optics, surface plasmon resonance, electrochemistry and near-field communication. The performances and advantages of these designs are introduced with their applications in healthcare diagnosis, environment monitoring and food evaluation with advances in micromanufacture, sensor technology and miniaturized electronics [70].

Using 3D-printing technology next to smartphone has provided the opportunity to turn any kind of smartphone into a portable luminometer to detect chemiluminescence derived from enzyme-coupled reactions. Roda et al. mentioned that the lactate oxidase was coupled with horseradish peroxidase for lactate determination in oral fluid and sweat. Lactate can be quantified in less than five minutes with detection limits of 4.5 mg/dL and 0.9 mg/dL in oral fluid and sweat, respectively. Devices based on smartphones offer an alternative to analytical performance with a cost-effective alternative for noninvasive lactate measurement. In the endurance sport and for monitoring lactic acidosis in critical-care patients [48], this can be used to detect and quantify the changes in the lactate with respect to the anaerobic threshold.

Recently in 2017, an improved design for a handheld smartphone-based spectrometer that works in both absorption and emission modes is proposed by de Oliveira et al. [71]. The device, named Spectrophone, comprises an embedded light source designed for absorption mode, a DVD for the diffraction grating, and a smartphone to process the image data acquired. User-friendly homemade software decomposes the pixels from shots of spectral images into their RGB and hue values. The spectrophone was applied to determine Fe^{2+} in medicine samples and Na^{+} in saline solution and natural water samples. No statistically significant differences were observed in comparison with commercial instruments with limits of quantification of 70 g/L and 60 g/L for absorption and emission modes, respectively.

Wang et al. [72] reported a multichannel smartphone spectrometer (MSS) as an optical biosensor that can simultaneously optical sense multiple samples with nanometer resolution. This optical sensor performed accurate and reliable spectral measurements by optical intensity changes at specific wavelength or optical spectral shifts. A custom smartphone Multiview App was developed to control the optical sensing parameters and to align each sample to the corresponding channel. The captured images were converted to the transmission spectra in the visible wavelength range from 400 to 700 nm with the high resolution of 0.2521 nm per pixel. The device was validated with the concentrations of protein and immunoassaying a type of human cancer biomarker, and the results showed that this MSS can achieve the comparative analysis detection limits, accuracy and sensitivity.

Vezzosi et al. employed a smartphone electrocardiograph (ECG) in evaluating heart rhythm and ECG measurements in dogs. A smartphone ECG tracing was recorded using a single-lead bipolar ECG recorder. Agreement between smartphone and standard ECG in the interpretation of tracings was evaluated. Sensitivity and specificity for the detection of arrhythmia were calculated for the smartphone. A perfect agreement between the smartphone and standard ECG was found in detecting bradycardia, tachycardia, ectopic beats and atrioventricular blocks. The smartphone ECG represents an additional tool in the diagnosis of arrhythmias in dogs, but it is not a substitute for a six-lead ECG. Arrhythmias identified by the smartphone ECG should be followed up with a standard ECG before making clinical decision [15].

Stedtfeld et al. developed the Gene-Z for the rapid and quantitative detection of genetic markers. The device is controlled by iPod Touch, to receive data and carried-out automated analysis and to report via Wi-Fi. This study presented data pertaining to performance of the device including sensitivity and reproducibility using genomic DNA from *Escherichia coli* and *Staphylococcus aureus* [73].

iHealth Lab Inc. has developed glucometers for smartphone, wireless smartphone glucometer, or mobile glucometer, which plugs directly into the smartphone's audio jack. iBGStar® from Sanofi and AliveCor®ECG for monitoring heart conditions are examples of commercial devices using the smartphone [1].

The method, adaptation of a smartphone's camera to function as a compact lens less microscope, is based on the shadow imaging technique where the sample is placed on the surface of the image sensor, which captures direct shadow images under illumination. The lens less imaging scheme allows for submicron resolution imaging over an ultrawide field of view. Image acquisition and reconstruction are performed on the device using a custom-built Android application, and constructing a stand-alone imaging device for field applications was presented by Lee and Yang [74].

Wireless chemical sensors are used as analytical devices in homeland defense, home-based healthcare and food logistics, and for that reason, Steinberg et al. [75] developed a portable potentiostat to perform mobile amperometric electrochemical measurements with wireless data transfer to other mobile devices. The developed device was compared with a model redox system, the reduction of hexacyanoferrate (III) and the commercial enzymatic blood glucose test strips.

A handheld and cost-effective cell phone-based colorimetric microplate reader uses a 3D-printed optomechanical attachment to hold and illuminate a 96-well plate using a light emitting diode (LED) array, which was proposed by Berg et al. [76]. The light is transmitted through each well and collected via 96 individual optical fibers. Then, the captured images are transmitted to the custom-designed app for processing using a machine learning algorithm, yielding diagnostic results, which are visualized by the user in same mobile application within less than 1 min per 96-well plate. This device was tested using FDA-approved mumps

IgG, measles IgG and herpes simplex virus IgG (HSV-1 and HSV-2) ELISA tests, working with 567 samples for training and 571 samples for blind. An accuracy of 99.6, 98.6, 99.4 and 99.4% for mumps, measles, HSV-1 and HSV-2 tests was achieved, respectively.

An optical fiber-based smartphone spectrometer incorporating an endoscopic fiber bundle is presented by Hossain et al. [77]. The endoscope allows transmission of the smartphone camera LED light to a sample, and the reflected spectra collected from a surface or interface is dispersed onto the camera CMOS using a reflecting diffraction grating. Spectral analysis of apples shows straightforward measurement of the pigments anthocyanins, carotenoid and chlorophyll, all of which decrease with increasing storage time.

Exploiting the abilities of the new technology, a mobile phone can serve the basic functions of a potentiostat in controlling an applied potential to oxidize electrochemiluminescence (ECL)-active molecules, while the resultant photonic signal is monitored using the camera. The excitation and detection processes are controlled by a software application which can also transmit the results via e-mail, which is the device presented by Delaney et al. [78].

3. Conclusion

Since the first smartphone was designed by IBM in 1993, several applications have been developed employing the smartphone as detector, scanner, quantifier and virtual interface, in medicine, environment, chemistry, food, biology, genetic, biotechnology, biomedical, instrumentation or computer sciences due to its portability, its price comparable with commercial devices or equipment and its own characteristics such as the camera, screen, to create mobile applications or the communication vias.

Detection of contaminants, pesticides, drugs, mycotoxins, vitamins, glucose, salivary cortisol, albumin, cholesterol, DNA molecules, proteins, bacterium, virus, cocaine, heavy metals, and amines in urine, water, blood, soil, and saliva are few examples of the immense quantity of the future applications, where the smartphone can be employed. This is motivated because the smartphone has become an indispensable device of our lives, increasing the number of users from day to day.

Author details

Diana Bueno Hernández[1,2]*, Jean Louis Marty[2] and Roberto Muñoz Guerrero[1]

*Address all correspondence to: dianaburh07@hotmail.com

1 Department of Electrical Engineering, Bioelectronics Section, CINVESTAV-IPN, Mexico

2 Université de Perpignan Via Domitia, BAE, Perpignan Cedex, France

References

[1] Roda A, Michelini E, Zangheri M, Di Fusco M, Calabria D, Simoni P. Smartphone-based biosensors: A critical review and perspectives. TrAC Trends in Analytical Chemistry. 2016;**79**:317-325

[2] Poushter J. Smartphone Ownership and Internet Usage Continues to Climb in Emerging Economies. Washington, DC: Pew Research Center; February 2016. pp. 1-44

[3] Ludwig SKJ, Zhu H, Phillips S, Shiledar A, Feng S, Tseng D, et al. The calibration of cell-phone camera-based colorimetric sensor array and its application in the determination of glucose in urine. Biosensors and Bioelectronics. 2015;**74**:1029-1037

[4] Lee S, Oncescu V, Mancuso M, Mehta S, Erickson D. A smartphone platform for the quantification of vitamin D levels. Lab on a Chip. 2014;**14**(8):1437-1442

[5] Aliev TA, Rzayeva NE, Sattarova UE. Robust correlation technology for online monitoring of changes in the state of the heart by means of laptops and smartphones. Biomedical Signal Processing and Control. 2017;**31**

[6] Messier R. Mobile computing. In: Ric M, Allan K, editors. Collaboration with Cloud Computing. Chapter 7. Massachusetts: Elsevier; 2014. pp. 115-133

[7] Mosa ASM, Yoo I, Sheets L. A systematic review of healthcare applications for smartphones. BMC Medical Informatics and Decision Making. 2012;**12**(67):1-31

[8] Rashvand HF, Hsiao KF. Smartphone intelligent applications: A brief review. Multimedia Systems. 2015;**21**:103-119

[9] Capecci M, Pepa L, Verdini F, Ceravolo MG. A smartphone-based architecture to detect and quantify freezing of gait in Parkinson's disease. Gait Posture. 2016;**50**:28-33

[10] Chun HJ, Park YM, Han YD, Jang YH, Yoon HC. Paper-based glucose biosensing system utilizing a smartphone as a signal reader. BioChip Journal. 2014;**8**(3):218-226

[11] Thom NK, Lewis GG, Yeung K, Phillips ST. Quantitative fluorescence assays using a self powered paper based microfluidic device and a camera equipped cellular phone. RSC Advances. 2014;**4**(3):1334-1340

[12] Chen A, Wang R, Bever CRS, Xing S, Hammock BD, Pan T. Smartphone-interfaced lab-on-a-chip devices for field-deployable enzyme-linked immunosorbent assay. Biomicrofluidics. 2014;**8**(6)

[13] Yu H, Tan Y, Cunningham BT. Smartphone fluorescence spectroscopy. Analytical Chemistry. 2014;**86**:8805-8813

[14] Wang Y, Liu X, Chen P, Tran TN, Zhang J, Chia WS, et al. Smartphone spectrometer for colorimetric biosensing. Analyst. 2016;**141**:3233-3238

[15] Vezzosi T, Buralli C, Marchesotti F, Porporato F, Tognetti R, Zini E, et al. Diagnostic accuracy of a smartphone electrocardiograph in dogs: Comparison with standard 6-lead electrocardiography. Veterinary Journal. 2016;**216**:33-37

[16] Kassianos AP, Emery JD, Murchie P, Walter FM. Smartphone applications for melanoma detection by community, patient and generalist clinician users: A review. British Journal of Dermatology. 2015;**172**:1507-1518

[17] Ribeiro N, Moreira L, Barros A, Almeida AM, Santos-Silva F. Guidelines for a cancer prevention smartphone application: A mixed-methods study. International Journal of Medical Informatics. 2016;**94**:134-142

[18] Figueiredo IN, Leal C, Pinto L, Bolito J, Lemos A. Smartphone application for emergency signal detection. Medical Engineering and Physics. 2016;**38**:1021-1027

[19] Zhu H, Isikman SO, Mudanyali O, Greenbaum A, Ozcan A, Wang J, et al. Optical imaging techniques for point-of-care diagnostics. Lab on a Chip. 2013;**13**(1):51-67

[20] Skandarajah A, Reber CD, Switz NA, Fletcher DA. Quantitative imaging with a mobile phone microscope. PLoS One. 2014;**9**(5):e96906

[21] Navruz I, Coskun AF, Wong J, Mohammad S, Tseng D, Nagi R, et al. Smart-phone based computational microscopy using multi-frame contact imaging on a fiber-optic array. Lab on a Chip. 2013;**13**:4015-4023

[22] Sicard C, Glen C, Aubie B, Wallace D, Jahanshahi-Anbuhi S, Pennings K, et al. Tools for water quality monitoring and mapping using paper based sensors and cell phones. Water Research. 2015;**70**:360-369

[23] Hussain I, Das M, Ahamad KU, Nath P. Water salinity detection using a smartphone. Sensors and Actuators B: Chemical. 2017;**239**:1042-1050

[24] Bueno D, Muñoz R, Marty JL. Fluorescence analyzer based on smartphone camera and wireless for detection of ochratoxin A. Sensors and Actuators B: Chemical. 2016;**232**:462-468

[25] Bueno D, Istamboulie G, Muñoz R, Marty JL. Determination of mycotoxins in food: A review of bioanalytical to analytical methods. Applied Spectroscopy Reviews. 2015;**50**(9):728-774

[26] Kwon L, Long KD, Wan Y, Yu H, Cunningham BT. Medical diagnostics with mobile devices: Comparison of intrinsic and extrinsic sensing. Biotechnology Advances. 2016;**34**:291-304

[27] Wargocki P, Deng W, Anwer A, Goldys E. Medically relevant assays with a simple smartphone and tablet based fluorescence detection system. Sensors. 2015;**15**:11653-11664

[28] Lee D, Chou WP, Yeh SH, Chen PJ, Chen PH. DNA detection using commercial mobile phones. Biosensors and Bioelectronics. 2011;**26**:4349-4354

[29] Guan L, Tian J, Cao R, Li M, Cai Z, Shen W. Barcode-like paper sensor for smartphone diagnostics: An application of blood typing. Analytical Chemistry. 2014;**86**:11362-11367

[30] Breslauer DN, Maamari RN, Switz NA, Lam WA, Fletcher DA. Mobile phone based clinical microscopy for global health applications. PLoS One. 2009;**4**(7):e6320

[31] Preechaburana P, Gonzalez MC, Suska A, Filippini D. Surface plasmon resonance chemical sensing on cell phones. Angewandte Chemie International Edition. 2012;**124**: 11753-11756

[32] Soni A, Jha SK. A paper strip based non-invasive glucose biosensor for salivary analysis. Biosensors and Bioelectronics. 2015;**67**:763-768

[33] Zangheri M, Cevenini L, Anfossi L, Baggiani C, Simoni P, Di Nardo F, et al. A simple and compact smartphone accessory for quantitative chemiluminescence-based lateral flow immunoassay for salivary cortisol detection. Biosensors and Bioelectronics. 2015;**64**:63-68

[34] Kim SC, Jalal UM, Im SB, Ko S, Shim JS. A smartphone-based optical platform for colorimetric analysis of microfluidic device. Sensors and Actuators B: Chemical. 2017;**239**:52-59

[35] Coskun AF, Nagi R, Sadeghi K, Phillips S, Ozcan A. Albumin testing in urine using a smart-phone. Lab on a Chip. 2013;**13**(21):4231-4238

[36] You DJ, Park TS, Yoon JY. Cell-phone-based measurement of TSH using Mie scatter optimized lateral flow assays. Biosensors and Bioelectronics. 2013;**40**(1):180-185

[37] Oncescu V, Mancuso M, Erickson D. Cholesterol testing on a smartphone. Lab on a Chip. 2014;**14**(4):759-763

[38] Wu Y, Boonloed A, Sleszynski N, Koesdjojo M, Armstrong C, Bracha S, et al. Clinical chemistry measurements with commercially available test slides on a smartphone platform: Colorimetric determination of glucose and urea. Clinica Chimica Acta. 2015;**448**:133-138

[39] Mancuso M, Cesarman E, Erickson D. Detection of Kaposi's sarcoma associated herpesvirus nucleic acids using a smartphone accessory. Lab on a Chip. 2014;**14**(19):3809-3816

[40] Wei Q, Luo W, Chiang S, Kappel T, Mejia C, Tseng D, et al. Imaging and sizing of single DNA molecules on a mobile phone. ACS Nano. 2014;**8**(12):12725-12733

[41] Priye A, Wong S, Bi Y, Carpio M, Chang J, Coen M, et al. Lab-on-a-drone: Toward pinpoint deployment of smartphone enabled nucleic acid-based diagnostics for mobile health care. Analytical Chemistry. 2016;**88**:4651-4660

[42] Khan SA, Smith GT, Seo F, Ellerbee AK. Label-free and non-contact optical biosensing of glucose with quantum dots. Biosensors and Bioelectronics. 2015;**64**:30-35

[43] Zhang D, Lu Y, Zhang Q, Liu L, Li S, Yao Y, et al. Protein detecting with smartphone-controlled electrochemical impedance spectroscopy for point-of-care applications. Sensors and Actuators B: Chemical. 2016;**222**:994-1002

[44] Oncescu V, O'Dell D, Erickson D. Smartphone based health accessory for colorimetric detection of biomarkers in sweat and saliva. Lab on a Chip. 2013;**13**(16):3232-3238

[45] Cho S, Park TS, Nahapetian TG, Yoon JY. Smartphone-based, sensitive µPAD detection of urinary tract infection and gonorrhea. Biosensors and Bioelectronics. 2015;**74**

[46] Libman D, Huang L. Chemistry on the go: Review of chemistry apps on smartphones. Journal of Chemical Education. 2013;**90**:320-325

[47] Salles MO, Meloni GN, de Araujo WR, Paixão TRLC. Explosive colorimetric discrimination using a smartphone, paper device and chemometrical approach. Analytical Methods. 2014;**6**(7):2047-2052

[48] Roda A, Guardigli M, Calabria D, Calabretta MM, Cevenini L, Michelini E. A 3D-printed device for a smartphone-based chemiluminescence biosensor for lactate in oral fluid and sweat. Analyst. 2014;**139**:6494-6501

[49] Li B, Li L, Guan A, Dong Q, Ruan K, Hu R, et al. A smartphone controlled handheld microfluidic liquid handling system. Lab on a Chip. 2014;**14**(20):4085-4092

[50] Smith JE, Griffin DK, Leny JK, Hagen JA, Chávez JL, Kelley-Loughnane N. Colorimetric detection with aptamer-gold nanoparticle conjugates coupled to an android-based color analysis application for use in the field. Talanta. 2014;**121**:247-255

[51] Chen L, Zhang C, Xing D. Paper-based bipolar electrode-electrochemiluminescence (BPE-ECL) device with battery energy supply and smartphone read-out: A handheld ECL system for biochemical analysis at the point-of-care level. Sensors and Actuators B: Chemical. 2016;**237**:308-317

[52] Wang H, Li YJ, Wei JF, Xu JR, Wang YH, Zheng GX. Paper-based three-dimensional microfluidic device for monitoring of heavy metals with a camera cell phone. Analytical and Bioanalytical Chemistry. 2014;**406**:2799-2807

[53] Rajendran VK, Bakthavathsalam P, Jaffar Ali BM. Smartphone based bacterial detection using biofunctionalized fluorescent nanoparticles. Microchimica Acta. 2014;**181**:1815-1821

[54] Park TS, Mccracken KE, Li W. Smartphone quantifies Salmonella from paper microfluidics. Lab on a Chip. 2013;**13**

[55] Bueno L, Meloni GN, Reddy SM, Paixão TRLC. Use of plastic-based analytical device, smartphone and chemometric tools to discriminate amines. RSC Advances. 2015;**5**(26)

[56] Liang P-S, Park TS, Yoon J-Y. Rapid and reagentless detection of microbial contamination within meat utilizing a smartphone-based biosensor. Scientific Reports. 2014;**4**(5953):1-8

[57] Yu L, Shi Z, Fang C, Zhang Y, Liu Y, Li C. Disposable lateral flow-through strip for smartphone-camera to quantitatively detect alkaline phosphatase activity in milk. Biosensors and Bioelectronics. 2015;**69**:307-315

[58] Monošík R, dos Santos VB, Angnes L. A simple paper-strip colorimetric method utilizing dehydrogenase enzymes for analysis of food components. Analytical Methods. 2015;**7**(19):8177-8184

[59] Ludwig SKJ, Zhu H, Phillips S, Shiledar A, Feng S, Tseng D, et al. Cellphone-based detection platform for rbST biomarker analysis in milk extracts using a microsphere fluorescence immunoassay. Analytical and Bioanalytical Chemistry. 2014;**406**:6857-6866

[60] Rico-Yuste A, González-Vallejo V, Benito-Peña E, de las Casas Engel T, Orellana G, Moreno-Bondi MC. Furfural determination with disposable polymer films and smartphone based colorimetry for beer freshness assessment. Analytical Chemistry. 2016; **88**:3959-3966

[61] Masawat P, Harfield A, Namwong A. An iPhone-based digital image colorimeter for detecting tetracycline in milk. Food Chemistry. 2015;**184**:23-29

[62] Andrade SIE, Lima MB, Barreto IS, Lyra WS, Almeida LF, Araújo MCU, et al. A digital image-based flow-batch analyzer for determining Al(III) and Cr(VI) in water. Microchemical Journal. 2013;**109**:106-111

[63] Gopinath SCB, Tang TH, Chen Y, Citartan M, Lakshmipriya T. Bacterial detection: From microscope to smartphone. Biosensors and Bioelectronics. 2014;**60**:332-342

[64] Wang Y, Li Y, Bao X, Han J, Xia J, Tian X, et al. A smartphone-based colorimetric reader coupled with a remote server for rapid on-site catechols analysis. Talanta. 2016;**160**:194-204

[65] Wei Q, Nagi R, Sadeghi K, Feng S, Yan E, Ki SJ, et al. Detection and spatial mapping of mercury contamination in water samples using a smart-phone. ACS Nano. 2014;**8**(2):1121-1129

[66] Lopez-Molinero A, Cubero VT, Irigoyen RD, Piazuelo DS. Feasibility of digital image colorimetry: Application for water calcium hardness determination. Talanta. 2013;**103**:236-244

[67] Jiang J, Wang X, Chao R, Ren Y, Hu C, Xu Z, et al. Smartphone based portable bacteria pre-concentrating microfluidic sensor and impedance sensing system. Sensors and Actuators B: Chemical. 2014;**193**:653-659

[68] Mei Q, Jing H, Li Y, Yisibashaer W, Chen J, Nan Li B, et al. Smartphone based visual and quantitative assays on upconversional paper sensor. Biosensors and Bioelectronics. 2016;**75**

[69] Choodum A, Kanatharana P, Wongniramaikul W, Nic Daeid N. Using the iPhone as a device for a rapid quantitative analysis of trinitrotoluene in soil. Talanta. 2013;**115**

[70] Zhang D, Liu Q. Biosensors and bioelectronics on smartphone for portable biochemical detection. Biosensors and Bioelectronics. 2016;**75**:273-284

[71] de Oliveira HJS, de Almeida PL, Sampaio BA, Fernandes JPA, Pessoa-Neto OD, de Lima EA, et al. A handheld smartphone-controlled spectrophotometer based on hue to wavelength conversion for molecular absorption and emission measurements. Sensors and Actuators B: Chemical. 2017;**238**:1084-1091

[72] Wang LJ, Chang YC, Sun R, Li L. A multichannel smartphone optical biosensor for high-throughput point-of-care diagnostics. Biosensors and Bioelectronics. 2017;**87**:686-692

[73] Stedtfeld RD, Tourlousse DM, Seyrig G, Stedtfeld TM, Kronlein M, Price S, et al. Gene-Z: A device for point of care genetic testing using a smartphone. Lab on a Chip. 2012;**12**:1454-1462

[74] Lee SA, Yang C. A smartphone-based chip-scale microscope using ambient illumination. Lab on a Chip. 2014;**14**(16):3056-3063

[75] Steinberg MD, Kassal P, Kereković I, Steinberg IM. A wireless potentiostat for mobile chemical sensing and biosensing. Talanta. 2015;**143**:178-183

[76] Berg B, Cortazar B, Tseng D, Ozkan H, Feng S, Wei Q, et al. Cellphone based hand held microplate reader for point of care testing of enzyme linked immunosorbent assays. ACS Nano. 2015;**9**(8):7857-7866

[77] Hossain MA, Canning J, Cook K, Jamalipour A. Optical fiber smartphone spectrometer. Optics Letters. 2016;**41**(10):2237-2240

[78] Delaney JL, Doeven EH, Harsant AJ, Hogan CF. Use of a mobile phone for potentiostatic control with low cost paper-based microfluidic sensors. Analytica Chimica Acta. 2013;**803**:123-127

Chapter 5

Semantic Annotation of Mobile Phone Data Using Machine Learning Algorithms

Feng Liu, JianXun Cui, Davy Janssens,
Geert Wets and Mario Cools

Additional information is available at the end of the chapter

http://dx.doi.org/10.5772/intechopen.70255

Abstract

Cell phone call location data has been utilized for the study of travel patterns, but the underlying activities that originate the movement are still at a less explored stage. Resulted from routine and automated features of decision-making processes, human activity and travel behaviour exhibit a high level of spatial-temporal periodicities as well as a certain order of the activities. In this chapter, a method has been developed based on these regularities, which predicts activities being conducted at call locations. The method includes four steps: a set of comprehensive variables is defined; feature selection techniques are applied; a group of state-of-the-art machine learning algorithms and an ensemble of the above algorithms are employed; an additional enhancement algorithm is designed. Using data gathered from natural communication of 80 users over a period of 1 year, the proposed method is evaluated. Based on the ensemble of the models, prediction accuracy of 69.7% was achieved. Using the enhancement algorithm, the performance obtained 7.6% improvement. The experimental results demonstrate the potential to annotate call locations based on the integration between machine learning algorithms and the characteristics of underlying activity and travel behaviour, contributing towards the semantic interpretation and application of the massive data.

Keywords: cell phone location annotation, activity and travel behaviour, machine learning algorithms, feature selection techniques, sequential information

1. Introduction

1.1. Problem statement

Nowadays, cell phones are frequently used as an attractive means for sensing human behaviour on a large scale. They provide a source of real and reliable data, enabling automatic monitoring call and travel behaviour of users. Studies have been conducted to discover statistical laws that govern the key dimensions of human travel, e.g. travel distance and time spent at different locations [1]. These studies provide a modelling framework capable of describing general features of human mobility.

However, despite the discovery of these general features, previous studies do not provide further insights into the motivation or activities behind the identified mobility features. In general, most of the current research on cell phone data has focused on spatial-temporal dimensions. The behavioural aspects associated with the mobility features, e.g. travel mode and activities being conducted at the locations, are still at a less studied stage. Due to privacy concerns, cell phone data provided by phone operation companies usually does not have contextual information, leading to a wide gap between the raw data and the semantic interpretation of the traces. If a method can be found which helps to bridge this gap, the potential applications of the semantically enriched phone data are immense. They include inferring people's travel motivations in activity-based transportation modelling, mining individual life styles and activity preferences in urban planning, and providing activity tailored services in the cell phone environment [2].

1.2. Related state of the art

Methods have been developed to derive activities being conducted at a location from global positioning systems (GPS)-based data or from multi-modal data recorded by cell phones. The GPS-based methods first decompose continuous GPS points into a chain of *stops,* where the individual stays for a minimum period of time conducting activities, and *moves* that are the points between two consecutive stops. The stops are then compared with a geographic map by matching them in space, and interesting places that are relevant to the studies are subsequently found. The GPS-based methods have received much attention during the past years [3], but are still faced with a number of limitations. (1) The data collection process is expensive in terms of battery consumption of GPS devices. (2) Linking a GPS trajectory to detailed geographic information on all interesting places in a study area needs a lot of computational work. (3) The methods are location-specific, and the quality of the annotation process depends on the study area, making the process not transferable to other areas. (4) The matched location alone may not disclose a particular reason of why an individual travels there. A person could go to a place (e.g. a shopping mall) with different purposes (e.g. working, shopping or having a lunch). (5) The matching of exact GPS positions raises privacy concerns, as some of the places visited by an individual may be highly privacy-sensitive.

Some of the above-described limitations have been addressed by the annotation process based on multi-modal data recorded from sensors equipped on cell phones [4]. This process is composed of two steps. In the first step, data from GPS and other sensors (e.g. Wi-Fi and accelerometer) is collected from each individual. The data is then clustered into a number of visit places, each of which is represented by an ID number rather than geographic positions of the cluster points. In the second step, the obtained places are annotated based on contextual information from the sensors and phone applications, as opposed to GPS data. In this process, various machine learning methods are proposed, and different sets of features are defined [5]. These studies achieved good prediction performance without the need of additional geographic information and GPS data. Nevertheless, while the machine learning methods eliminate the need for a map, this entire annotation process still partly relies on GPS data for the identification of visit places in the first step. Thus, this process as a whole does not fully address the privacy concern. On top of that, while these studies mainly focus on selecting efficient classification models and relevant features, none of them have conducted post-processing analysis to examine how the predicted results are consistent with the sequential information that is embedded in daily activity and travel sequences. In-depth examination into the prediction errors is also lacking in these studies.

1.3. Research contributions

Extending the current research on annotating people's movement traces, our study proposes a new approach. The method utilizes data collected from simple cell phones, and it combines machine learning methods with the characteristics of underlying activity and travel behaviour that originates the traces. It has the following advantages over the existing studies. (1) The method is based on spatial-temporal regularities as well as sequential information intrinsic to human activity and travel behaviour. (2) It does not depend on additional sensor data and map information, reducing data collection costs and increasing transferability. (3) An enhancement algorithm has been developed to improve the prediction results by machine learning methods. (4) A set of extensive experiments and in-depth examination into the classification errors have been conducted. (5) Compared to GPS points, the wide coverage of a cell ID allows the process to reduce privacy concerns considerably.

The rest of this paper is organized as follows. Section 2 introduces the cell phone data and Section 3 elaborates on the annotation process. Experiments are conducted in Section 4 and examination into the experiment results is carried out in Section 5. Finally, Section 6 ends this chapter with major conclusions and discussions for future research.

2. Data

The cell phone data is composed of full mobile communication patterns of 80 users over a period of 1 year, collected by a European phone company for billing and operational purposes. The data records the location and time when each user performs a call activity, including initiating or receiving a voice call or message. The locations are represented with cell IDs, each of which has a coverage ranging from a few hundred square metres in cities to a few thousand in rural areas. The users along with their phone numbers and the corresponding cell IDs are all anonymized. **Table 1** illustrates typical call records of an individual identified as '10027534' on a day.

User ID	Cell ID	Time	Duration	Call type	Direction
10027534	10163	10:18	12	Voice call	Outgoing
10027534	10269	12:40	0	Message	Incoming

[a]The columns, respectively, denote the user, cell ID, time and duration (in minutes) of the call, the call type including 'voice call' and 'message' and the direction including 'incoming', 'outgoing' and 'missed calls'.

Table 1. Call records of a user.[a]

Among all the users, 9132 distinct call locations were detected and 259 (2.8% of the total identified locations) were labelled with activities conducted at these places. These labelled locations are used as the ground-truth data for training and validating our models. Activities are divided into five types, including 'work/school', 'home', 'social visit', 'leisure' and 'non-work obligatory', accounting for 30, 29, 15, 14 and 12% of the training data, respectively. The type of 'work/school' represents all work- or school-related activities outdoors; while 'home' accommodates all time spending at home. 'Social visit' refers to all visit activities, 'leisure' includes recreational activities outside home, e.g. sports and eating/drinking, and 'non-work obligatory' consists of activities like bringing/getting people, shopping and personalized services. If activities in multiple types are executed in the same location for a particular individual, the most frequent activity is selected, such that each location is uniquely linked to an activity type for the individual.

3. Methodology

3.1. Overview of the approach

The approach incorporates basic knowledge about human activity and travel decision-making processes and their resultant activity and travel behaviour. As Liu et al. [6] underlined, human activity and travel decision-making processes demonstrate routine and automated features. People do not generally schedule their activities on a daily basis; but rather depend on fixed routines or scripts executed during the day without much alteration. This leads to a high level of spatial-temporal regularities in activity and travel behaviour as well as a certain sequential order of the activities [6]. The spatial-temporal recurrences of the locations can be adequately reflected in the movement traces of cell phone users through a long period of call records. In addition, the spatial-temporal constraints of locations, stemming from the characteristics of various activities, which are performed in their own daily, weekly or monthly rhythms, can thus suggest the possible activities carried out at the locations. This enables the annotation for the third dimension, i.e. travel motives (activities). Furthermore, evidence also suggests that activity and travel behaviour differs across various time periods of a day, between weekdays and weekends, and between normal days and holidays [7].

The method consists of four major steps. (1) A set of variables characterizing call locations in the spatial-temporal dimensions is defined. (2) Feature selection techniques are applied to choose the most effective variables. (3) Upon the obtained variables, a set of classification models and an additional ensemble method to combine these prediction results are employed. (4) An enhancement algorithm is developed to improve the annotation performance based on sequential constraints of the activities.

3.2. Variable definition

For each user, all distinct locations, where the person has performed at least a call activity during the entire data collection period, are extracted. Let N as the total number of these locations. At each location Loc_i $(i = 1 \ldots N)$, a set of variables is defined from two perspectives, including the *call behaviour* and the underlying *travel behaviour*. The *call behaviour* defines the variables that are directly related to call communication activities. Most of the variables are also used in the multi-modal data annotation process, as described in Section 1. The *travel behaviour*, however, approximates the spatial-temporal features of a location. The difference between these two perspectives can be illustrated by two groups of major variables. The first group includes the call frequency CFreqR and visit frequency VFreqR. CFreqR depicts how often calls are made at a location; by contrast, VFreqR reveals how often the location is reached, irrespective of the number of calls that are made at each visit. The second is the call duration CDur and visit duration VDur. CDur describes the duration of the call; while VDur is defined as the time interval between the first and last calls at the location. Apart from the different perspectives, all the variables are also divided based on spatial-temporal factors, including spatial repetition, temporal periodicity, day types and day segments. All the variables are listed in **Table 2**.

In terms of day segments, different definitions of time periods have been adopted, depending on the context of the study area [8]. Instead of making such an a priori assumption, a method that is proposed in this study estimates the splitting points of the day from empirical data. The resultant splitting points delimit the largest difference in the distribution of various activity types across these time intervals. Specifically, the segment process starts with a full day of 24 hours, and each hour is examined independently. An hour under investigation divides the day into two time intervals, e.g. 0–10 am and 10 am to 24 pm at 10 am. A contingency table is then constructed, in which these two time intervals and the five activity types are the row and column variables, respectively. The frequencies of the aggregated observations from the labelled call locations that fall into the corresponding time intervals and activity classes are the cell values. A chi-square statistics is subsequently calculated for this table. After chi-square statistics is obtained for each of the 24 hours, the hour with the largest statistics is chosen as the first splitting point, denoted as S_1. This point divides the day into two intervals between 0 and S_1 as well as between S_1 and 24. This process is repeated for each of the latest formed intervals, until further splitting does not generate substantial difference or until a pre-specified number of intervals is reached.

3.3. Feature selection

Due to the small size of the training dataset, particularly relative to the large number of defined variables, over-fitting is a potential problem. To address this issue, feature selection techniques are employed in order to decrease the number of predictors actually utilized by the

Travel behaviour

Spatial repetition. **(1) VFreqR**: the visit frequency at the location divided by the total visit frequencies to all locations by the individual.

Temporal variability. **(1) TotVDurR**: the total duration of all the visits to the location divided by the duration of visits to all locations by the individual. **(2) [Ear/Lat]VTime**: the earliest and latest call time of all calls at the location. **(3) AveV [StartT/ EndT], VarV[StartT/EndT]**: the average and variance of the first and last call time over all visits at the location. **(4) [Longest/Ave/Var]VDur**: the longest and average duration of all visits to the location, and the variance of the duration.

Day type. **(1) VFreqR[Week/Weekend/Sun/Sat/Hol],TotVDurR**

[Week/Weekend/Sun/Sat/Hol]: 'VFreqR' and 'TotVDurR' at weekdays, weekend, Sunday, Saturday, or public holidays.

Day segment. **(1) VFreqR[1/.../m], TotVDurR[1/.../m]:** 'VFreqR' and 'TotVDurR' are segmented during different time periods of a day.

Call behaviour

Spatial repetition. **(1) CFreqR**: the call frequency at the location divided by the total call frequencies at all locations by the individual. **(2) [VoiC/Mes]FreqR**: 'CFreqR' is segmented between voice calls and messages. **(3) [Inc/Mis/Out]CFreqR**: 'VoiCFreqR' is divided into incoming, missed and outgoing **(4) [Inc/Out]MesFreqR**: 'MesFreqR' is divided into incoming and outgoing.

Temporal variability. **(1) TotCDur'**: the total call duration of all calls at the location by the individual. **(2) CInt[Max/Ave]**: the maximum and average time interval between 2 consecutive calls at the location. **(3) [Ave/Var]CTime**: the average and variance of call time of all calls at the location. **(4) [Longest/Ave/Var]CDur'**: the longest, average and variance of duration of all calls at the location.

Day type. **(1) CFreqR[Week/Weekend/Sun/Sat/Hol], TotCDur'R**

[Week/Weekend/Sun/Sat/Hol],VoiCFreqR[Week/Weekend/Sun/Sat/Hol], MesFreqR[Week/Weekend/Sun/Sat/Hol]: 'CFreqR', 'TotCDur''. 'VoiCFreqR' and 'MesFreqR' at weekdays, weekend, Sunday, Saturday, or public holidays.

Day segment. **(1) CFreqR[1/ .../ m], TotCDur'R[1/ .../ m], VoiCFreqR[1/ .../ m], MesFreqR[1/ .../ m]:** 'CFreqR', 'TotCDur'', 'VoiCFreqR' and 'MesFreqR' are segmented during different time periods of a day.

[a]The symbol [] denotes different variables, e.g. [Ear/Lat]VTime for variables 'EarVTime' and 'LatVTime'. Each day is divided into m segments, and m is decided by the method described as follows.

Table 2. Variable definition.[a]

classification models. Two methods including wrapper [9] and filter [10], which have shown effectiveness in the multi-modal data annotation process, are chosen for feature selection. Wrapper searches for an optimal feature subset using the classification model itself. In contrast, filter examines each feature separately and selects the feature that has high correlation with the target variable, but low relation with the features that have already been chosen.

3.4. Machine learning

A group of state-of-the-art machine learning algorithms, including decision trees (DTs) [11], random forests (RF) [12], multinomial logistic regression (MNL) [13] and multiclass support vector machines (SVMs)[14], are employed. These algorithms have demonstrated comparative performance for multi-category classification problems. These methods mainly differ in terms of the way the classification question is formulated, the learning function and the solution to deciding the optimal function parameters. As each learning algorithm has its strength and weakness, it is often challengeable to identify a single algorithm that performs best for a particular classification problem [15]. Thus, in this study, a fusion process is developed, which integrates the results of these algorithms, in order to utilize the strength of one while complementing the limitation of another. In this process, the four individual model prediction results (i.e. the probabilities of different possible activity types) for each call location are used as predictors, and the observed activity types are still as the dependent variable. The correlation between these predictors and the observed activity types can be built again by a classification model.

3.5. The enhancement algorithm

While machine learning methods provide an effective solution to annotating each single location, they disregard the activity orders and transitions embedded in daily activity and travel sequences. When the annotated locations on a day are linked according to the temporal order, they should follow a certain sequential constraint. The interdependencies of daily activities have been considered as a crucial factor in activity and travel decision making, as discussed in Section 3.1. By considering sequential information, the activity locations that are accessed by an individual on a day are viewed and tackled as a whole, rather than isolated participation in activities.

The enhancement algorithm takes the preliminary inference results as well as the sequential knowledge as inputs and aims to improve the prediction. The method is composed of two components: transition probability-based enhancement and prior probability-based enhancement. **Figure 1** illustrates how the prediction is improved using a daily location sequence of a user.

According to the training data of the user, he/she has conducted the chain of activities of 'work-social visit-work' at the respective call time on a day. But the prediction from the classification models is 'work-non-work obligatory-work'. A prediction error occurs at the second location. In this case, if a location (e.g. the second location) has a prediction probability P (0.443) smaller than a threshold T_1 (0.72 in our case study), it is assumed that the location is likely to be wrongly annotated. The enhancement algorithm is then applied to the false location to improve its prediction in the following steps. (1) If there is an additional location adjacent to the false one (including backwards and forwards) in the predicted sequence for that day and if this location has P larger than a threshold T_2 (0.9), it is considered as possibly correct prediction. The additional location is thus used to fix the prediction of the false one, using the

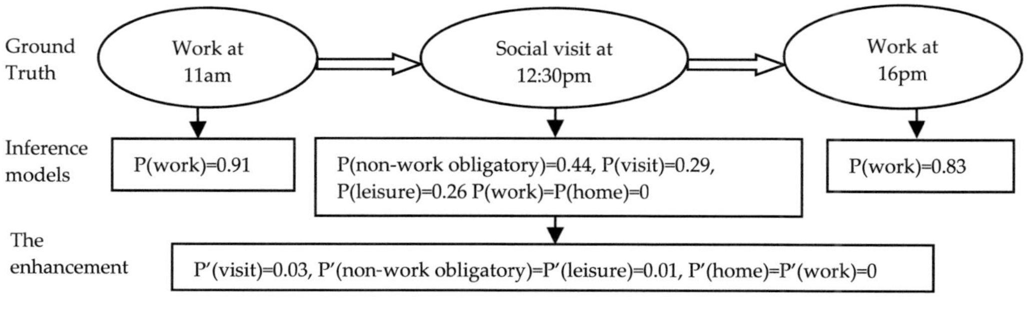

Figure 1. A daily call location sequence.

transition probability-based enhancement. (2) Otherwise, if no other locations in the neighbouring areas are predicted with a high probability, the *prior probability-based enhancement* method is employed to increase the prediction accuracy based on the call time at the false location. After recalculation, the activity type with the largest enhancement probability P' is chosen as the annotation result of the false location on that particular day. As a location may be repeatedly visited on multiple days, the multiple days' enhancement results are integrated by majority voting rules as the final annotation for the location. Under the appropriate parameters T_1 and T_2, the false prediction is likely to be corrected while accurate inference results are maintained. **Figure 2** demonstrates the details of the enhancement process.

3.5.1. *Transition probability-based enhancement*

The sequential information is represented in a 5 × 5 transition probability matrix between different activities. Let a_i and a_j (a_i, a_j = 1,...5) as the activities performed at the previous location i and current location j, respectively; $Tr(a_j|a_i)$ as the transition probability from a_i to a_j, calculated from the training data as follows:

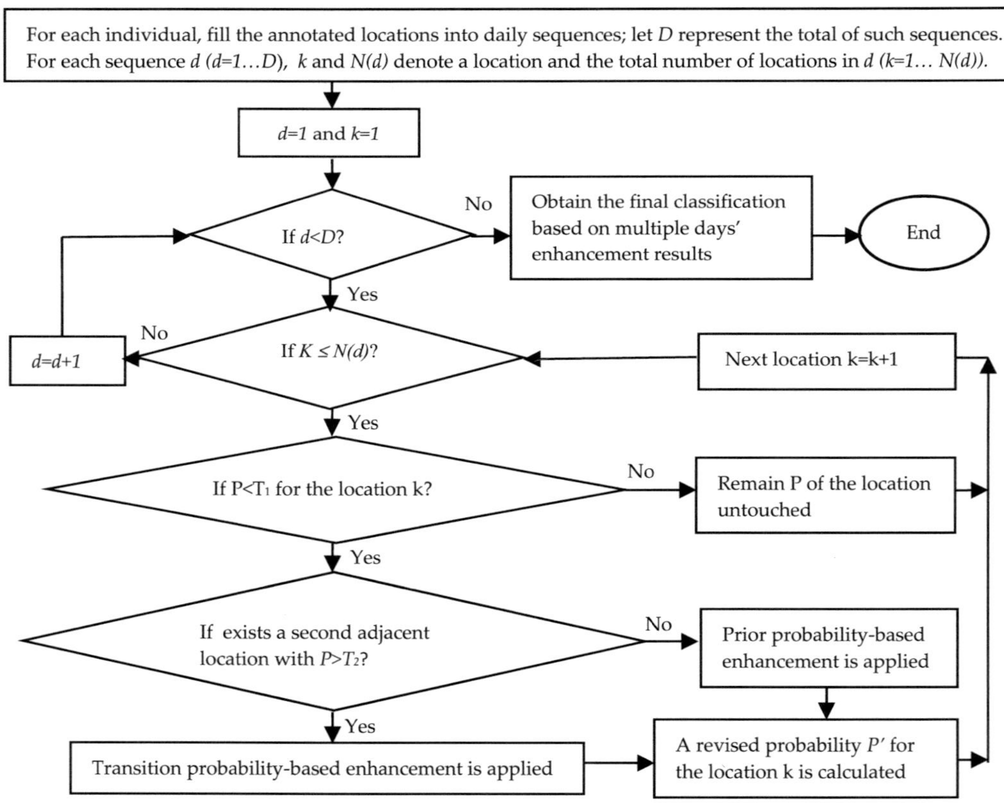

Figure 2. The enhancement algorithm.

$$Tr(a_j|a_i) = \frac{F(a_j|a_i)}{\sum_{a_k=1}^{5} F(a_k|a_i)} \quad (1)$$

$F(a_j | a_i)$ is the frequency of a_j followed by a_i. The probability of the location j being annotated as a_j conditioned by a_i at the previous location i can be recalculated as $P^0(a_j | X)$ according to Eq. (2).

$$P^0(a_j|X) = P(a_j|X) \times Tr(a_j|a_i) \quad (2)$$

$P(a_j | X)$ is the result of the classification model. It is noted that $P^0(a_j | X)$ is biased towards frequently visited locations, e.g. home and work/school places, as transitions to these places are more likely than to other less visited locations. Consequently, most of the locations under Eq. (2) will be redirected to these two activity types. To overcome this, $Tr(a_j | a_i)$ is divided by the frequency of a_j, resulting in the probability $Qr(a_j | a_i)$.

$$Qr(a_j|a_i) = \frac{F(a_j|a_i)}{\sum_{a_k=1}^{5} F(a_k|a_i) \times \sum_{a_k=1}^{5} F(a_j|a_k)} \quad (3)$$

$P^0(a_j | X)$ can be revised as $P'(a_j | X)$.

$$P'(a_j|X) = P(a_j|X) \times Qr(a_j|a_i) \quad (4)$$

In the user's case, as shown in **Figure 1**, since the transition probability Qr from work to non-work obligatory activities is very small, after the enhancement, $P'(non-work-obligatory)$ (0.008) drops behind $P'(visit)$ (0.033), we get the visit activity as the revised annotation.

3.5.2. Prior probability-based enhancement

The above-described transition probability-based enhancement involves at least two locations, which are adjacent in time, and one of which has a prediction probability larger than T_2. However, such daily trajectories derived from the classification models are not always available for each day. For example, one of the neighbouring locations has a probability smaller than T_2. Or, in the case where people may stay at a location (e.g. home) during an entire day, engaging only in a single (home) activity. This is particularly true with cell phone data. People may not make calls when travelling to an activity location, resulting in the daily movement traces not being fully revealed by their call data. In these cases, we utilize the typical activity and travel behaviour at different time of a day through the prior probability distribution of the activity a_j at different call time t, i.e. $P(a_j | t)$. By applying Bayesian methods, we compute the posterior probability of a_j based on X and t, i.e. $P'(a_j | X, t)$. This probability can be computed as follows, with the assumption that X is independent of t.

$$P'(a_j|X,t) = \frac{P(a_j,X,t)}{P(X,t)} = \frac{P(X,t|a_j) \times P(a_j)}{P(X) \times P(t)}$$
$$= \frac{P(a_j|X) \times P(X)}{P(a_j)} \times \frac{P(a_j|t) \times P(t)}{P(a_j)} \times \frac{P(a_j)}{P(X) \times P(t)} \tag{5}$$
$$= \frac{P(a_j|X) \times P(a_j|t)}{P(a_j)}$$

$P(a_j|X)$ is the output of the classification model, i.e. the probability of a_j performed at the location j conditioned on the previously defined variables X. When $P(a_j|X)$ is compared with the new probability $P'(a_j|X,t)$, since t is added in the conditional part of P', the new probability is more discriminative and informative than P.

$P(a_j|t)$ and $P(a_j)$ can be derived from the training data as follows:

$$P(a_j|t) = \frac{F(a_j|t)}{\sum_{a_k=1}^{5} F(a_k|t)}$$
$$P(a_j) = \frac{F(a_j)}{\sum_{a_k=1}^{5} F(a_k)} \tag{6}$$

Here, $F(a_j|t)$ refers as the occurrences of a_j at t and $F(a_j)$ refers as the occurrences of a_j at all time. It should be noted that from the theoretic perspective, the above enhancement process has two weak assumptions. One is the replacement of $P(a_j|X)$ with the result of the classification model and the other concerns the hypothesis of the independence between X and t. Nevertheless, based on Eq. (5), the preliminary prediction probability is complemented with the prior probability distribution.

4. Case study

In this section, adopting the proposed method and using the cell phone data described in Section 2, a set of experiments is presented. The results of these experiments are discussed and the performance of the annotation process is evaluated.

4.1. Day segments

Table 3 lists the optimal points for each of the intervals, based on the method described in Section 3.2. The first splitting point over an entire day was found at 9 am, generating two intervals of 0–9 am and 9 am to 24 pm. This process was iterated for each of the two newly obtained intervals. If the largest chi-square value over all potential points of an interval was lower than a predefined threshold, i.e. 200 in this experiment, this search stops.

Current interval	[0,24]	[0,9]	[9,24]	[9,19]	[19,24]	[9, 14]	[14,19]
S	9 am	7 am	19 pm	14 pm	20 pm	10 am	16 pm
Chi-square	3302	139	1603	855	75	194	30
If split?	Yes	No	Yes	Yes	No	No	No
New intervals	[0,9], [9, 24]	X	[9,19], [19,24]	[9, 14], [14,19]	X	X	X
Order	1	5	2	3	6	4	7

[a]The rows, respectively, denote the current interval (hour) under investigation, the optimal splitting point *S*, the chi-square value, the decision on whether or not the interval is split (if it is 'Yes' then two new intervals are formed and if it is 'No' then the symbol 'X' is used), and the order of the optimal points according to the chi-square values.

Table 3. The optimal points of a day.[a]

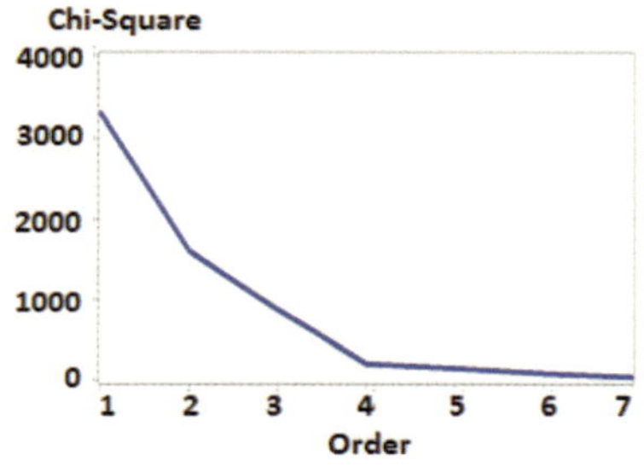

Figure 3. The evolution of chi-square statistics of the optimal points.

Figure 3 further shows the evolution of the chi-square statistics, in which the first 3 orders yield much higher values than the remaining ones. From the fourth order on, the statistics starts to decline sharply. Thus, the first 3 optimal points were extracted and 4 time periods were generated including 0–8:59 am, 9–13:59 am, 14–18:59 pm and 19–23.59 pm. After each day was segmented into the four periods, all the variables defined in **Table 2** were obtained and used as candidates for subsequent feature selection and machine learning. Weka, an open-source Java application consisting of a collection of machine learning algorithms for data mining tasks [16], was used for the implementation.

4.2. Results of individual classification models

The original training dataset is randomly divided into 10 subsets. In each model run, one of these subsets is used as the validation data and the remaining subsets combined as the training data. The number of correctly annotated locations in the validation subset is denoted as $C_i(i = 1\ldots10)$. Let *Num* as the total number of locations in the training dataset; the *prediction accuracy* can be defined as follows:

$$\text{Accuracy} = \frac{\sum_{i=1}^{10} C_i}{\text{Num}} \tag{7}$$

The individual classification models are built on the features of locations drawn from the perspectives of both travel and call behaviour as well as on the features profiling only call behaviour, respectively. In addition, the models are also run separately on all candidate variables as well as on the variable subsets that are chosen by filter or wrapper. The prediction results with the best parameter setting in each case are presented in **Table 4**.

From the prediction results, the following observations can be drawn. (1) The models running on a subset of variables perform better than those operating on all predictors. The average improvement is 0.85% for wrapper and 2.13% for filter. This demonstrates the importance of feature selection techniques in dealing with a large number of predictors relative to a small training set. (2) There are no general conclusions on which feature selection methods are better, depending on specific classification models. SVM performs better with filter, DT and RF do not show much difference between these two feature selection techniques, while MNL gains remarkable improvement of 4.8% with wrapper. (3) When the different models are compared, it is noted that MNL produces the best results with 68.98% accuracy. This is followed by accuracy of 66.06% from RF, 65.69% from SVM and 60.95% from DT. (4) Variation is also exhibited between the variables drawn from different perspectives. In most cases, the prediction accuracy derived from the combination of both travel and call behaviour is higher than that from solely call behaviour. The average accuracy increases by 2.96 and 1.20% for filter and wrapper, and 2.09% for all variables included. This underlines the added value of the variables built based on underlying activity and travel behaviour.

Apart from different model performance, the feature selection techniques combined with various classification models also yield divergent optimal subsets of features. Eight variables are picked up by the multiple selection processes and they are regarded as important predictors, including VFreqRWeek, TotVDurRSun, VarVEndT, VarVStartT and AveVEndT describing activity and travel behaviour, and AveCallTime, IncMesFreqR and MesFreqR3 related to only call behaviour.

Classification models	**DT**	**RF**	**MNL**	**SVM-poly**	**SVM- RBF**
Parameters	$N = 4$	$N = 0$	$C = 1$	$c = 100$, degree = 1	$c = 100$, Gamma = 0.01
Travel and call behaviour					
Filter	**60.95**	65.33	64.23	63.50	**65.69**
Wrapper	**1.1.** 60.58	**1.2. 66.06**	**1.3. 68.98**	**1.4.** 59.26	**1.5.** 56.57
1.6. All Variables	**1.7.** 59.12	**1.8.** 64.60	**1.9.** 63.50	56.93	**1.10.** 59.85
Call behaviour					
Filter	58.76	62.77	62.77	59.85	60.58
Wrapper	59.85	63.50	65.69	59.49	58.39
All variables	56.57	62.04	60.58	57.30	59.85

[a]The highest prediction accuracy for each model is in bold.

Table 4. Prediction accuracy of the individual classification models (%).[a]

4.3. Results of fusion models

In this fusion process, the four individual classification models are, respectively, employed as the fusion models to predict the activity types, while the results from each of the classifiers with the best parameter performance shown in **Table 4** are used as the predictors. The prediction with the two best performances for each fusion model is presented in **Table 5**. The results reveal that a fusion model does not necessarily outperform the individual models; the performance depends on the choice of the selected individual classifiers as the predictors. For instance, MNL obtains 68.98% accuracy as an individual classifier, while it achieves 69.71% when used as the fusion model built on the integration of all the four individual models' results. However, the accuracy drops to 61.68% when only DT and SVM-RBF are employed as the predictors.

4.4. Enhancement algorithm

4.4.1. Transition matrix

Similar to the temporal variables, the transition matrix is also built for weekdays, weekend and holidays separately as well as for different periods of a day. The identification of optimal cutting points for the matrix is the same as the previously described method, except the time intervals. For each potential dividing point, two intervals but three scenarios are obtained depending on the time of the two concerned activities in the transition. The first and second scenarios occur when both activities take place in the first interval or in the second. The third scenario is when the first activity takes place in the first interval and second activity in the second interval. Given the small size of the training set, only the first significant cutting point was identified, which is 18 pm. Under this time division, the largest difference in the distribution of activity transitions is among the three scenarios: transitions within 0–17:59 pm or 18–23:59 pm, and transitions from 0–17:59 pm to 18–23:59 pm. **Table 6** shows the transition matrix in the first scenario during weekdays.

Predictor	DT	RF	MNL	SVM - RBF	Accuracy
Fusion models					
DT			X	X	69.71
DT	X		X		67.15
RF		X	X		68.98
RF			X	X	68.24
MNL	X	X	X	X	69.71
MNL		X	X		68.98
SVM-RBF	X	X	X	X	67.52
SVM-RBF		X		X	67.15

[a]The rows represent the fusion models, and the columns include the individual classifiers and the prediction accuracy. X indicates the corresponding individual models being chosen as the predictors.

Table 5. Prediction accuracy of fusion models (%).[a]

Transition probability	Activity type	Home	Work/school	Non-work	Social visit	Leisure
Tr	**Home**	0.008	**0.546**	**0.700**	0.197	**0.797**
	Work/school	**0.883**	0.328	0.300	**0.701**	0.153
	Non-work	0.032	0.010	0.000	0.000	0.000
	Social visit	0.017	0.081	0.000	0.080	0.051
	Leisure	0.061	0.036	0.000	0.022	0.000
Transition probability	**Activity type**	**Home**	**Work/school**	**Non-work**	**Social visit**	**Leisure**
Qr	**Home**	0.002	**0.159**	**0.204**	0.057	**0.232**
	Work/school	0.060	0.022	0.020	0.047	0.010
	Non-work	**0.066**	0.019	0.000	0.000	0.000
	Social visit	0.023	0.114	0.000	**0.113**	0.072
	Leisure	0.059	0.035	0.000	0.021	0.000

[a]The row and column represent the current and previous activities respectively; the maximum probability for each column is in bold.

Table 6. Transition matrix.[a]

As expected, for the probability $Tr(a_j | a_i)$, the highest values are dominated by the transitions to either home or work/school activities. With $Qr(a_j | a_i)$, however, the dominance of these two activities is reduced by their high frequencies, and transitions to other less represented activities are exposed. This can be manifested by the high transitions from home to non-work activities and from social visit to second social visit locations.

4.4.2. *Activity distribution at different time*

The activity distribution is also differentiated between weekdays, weekend and holidays. The weekday distribution at each hour $P(a_j | t)$ is shown in **Figure 4(a)** and the distribution of the

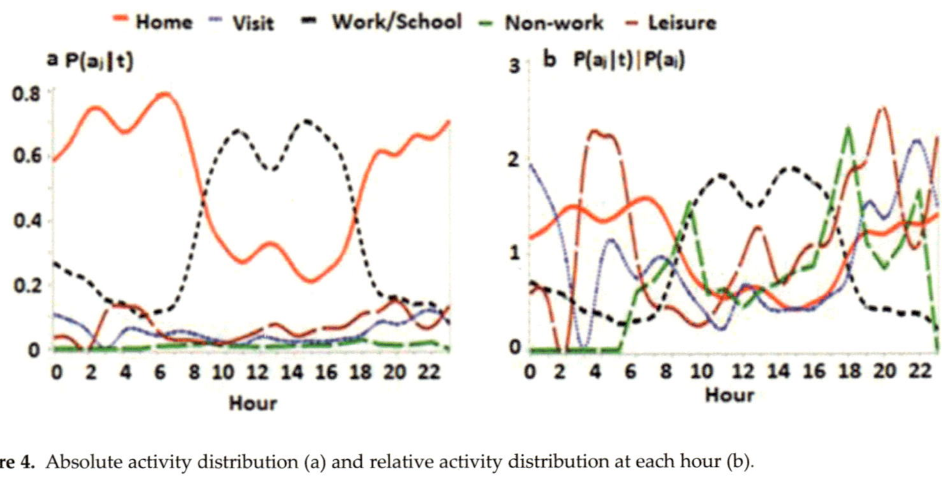

Figure 4. Absolute activity distribution (a) and relative activity distribution at each hour (b).

ratio between $P(a_j|t)$ and the overall probability of the activity $P(a_j)$ is depicted in **Figure 4(b)**. These two distributions show remarkable deviation: in **Figure 4(a)** either home or work/school types dominate the activities, whereas in **Figure 4(b)** the most likely activity shifts across various types as the day unfolds.

4.4.3. Selection of T_1 and T_2

Based on the previous results, two fusion models, including MNL built on all the four individual classifiers and RF on the combination between this model and MNL, are selected for the enhancement algorithm. To decide the threshold T_1, the correlation is examined between different values of T_1 and the prediction rates of the fusion models, as shown in **Figure 5**. It is observed that for both models, when T_1 is below the crossing point of 0.72 in **Figure 5(a)** and 0.8 in **Figure 5(b)**, the number of false prediction is higher than that of the correct one. Thus, 0.72 and 0.8 are selected as T_1 for MNL and RF, respectively. T_2 is set as 0.9, above which the prediction rate is 69.7 and 66.4% for these two models.

4.4.4. Enhancement results

Table 7 presents the prediction results by the enhancement algorithm (in the column 'After'), along with the results before the enhancement (in the column 'Before') as well as the difference between these two prediction results (in the column 'Difference'). Overall improvement of 4.4 and 7.6% for MNL and RF is achieved. The examination into the results across various activities discloses that the enhancement algorithm particularly performs better on less representative activity types, e.g. non-work obligatory, social visit and leisure activities. This could be originated from the fact that the machine learning algorithms usually favour majority types if the prediction accuracy is used as the evaluation criterion, while the enhancement algorithm puts equal weights on all activity types of the dependent variable (call locations).

The effectiveness of each of the two enhancement methods is also investigated, by running the RF fusion model using each of these methods independently to revise a weak prediction result.

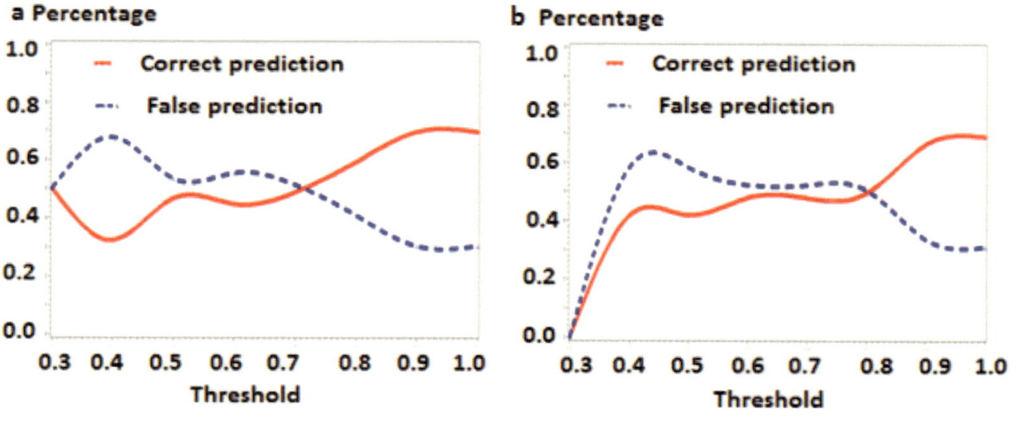

Figure 5. Relation between the prediction and the probabilities from MNL (a) and RF (b) fusion models, respectively.

Fusion model	MNL			RF		
Enhancement	**Before**	**After**	**Difference**	**Before**	**After**	**Difference**
Home	91.3	91.3	0	91.3	91.3	0
Work/school	80.9	82.0	1.1	74.2	79.8	5.6
Non-work	37.5	59.3	21.8	53.1	78.1	25.0
Visit	47.4	55.3	7.9	52.6	60.5	7.9
Leisure	45.7	48.6	2.9	37.1	51.4	14.3
Overall accuracy	69.7	74.1	4.4	69.0	76.6	7.6

Table 7. Prediction result comparison between before the enhancement algorithm and after that (%).

The prediction rates of 73.7 and 75.2% were obtained for the transition probability-based and prior probability-based enhancement methods, respectively. Due to the small size of the training set, many locations are labelled as one single known activity of a day, the sequential information is thus not available on these days. With a large dataset, the transition matrix would better represent typical activity and travel behaviour of users. This would lead to the transition probability-based method and the enhancement algorithm as a whole bringing greater improvement over the current experimental results.

5. Analysis on the prediction results

Table 8 presents the annotation results by the RF fusion model with the enhancement algorithm, showing a large variation in the prediction accuracy across different activity types. Home, work/school and non-work obligatory activities are better predictable, with the accuracy of 91.3, 79.8 and 78.1%, respectively. Social visit activities show a middle level of predictability of 60.5%. By contrast, leisure activities are only 51.4% recognizable. Overall, prediction accuracy of 76.6% is achieved. Despite the promising results, misclassification exists for each of the activity types, prompting for further examination into the potential reasons for the errors.

(1) Home. Homes are featured with high visit frequencies and spatial-temporal regularities. However, seven homes are misidentified, of which five have lower visit frequencies than 10% on weekdays, i.e. less than 1 in 10 trips on weekdays ending at home. The unusually less visited homes could be due to the fact that the corresponding users spend less time at home and/or they make fewer calls than expected at home. This results in the home visit frequencies less represented by their call records. Alternatively, some of the misclassified locations can be a second home for users who already have a home at different locations. Two in these five users have two labelled homes. While their second homes are occasionally accessed, their main homes are routinely visited and correctly annotated. (2) Work/school. Like homes, work/school locations are also characterized by a high level of routine visits, but these two types differ regarding the time of the visits. While most of the trips to homes are at night and weekends, trips to offices or schools occur during the daytime on weekdays. Of all the work/school

Annotated activity	Original activity				
	Home	Work/school	Non-work	Social visit	Leisure
Home	91.3	10.1	3.1	15.8	2.8
Work/school	1.2	79.8	9.3	7.9	14.3
Non-work	2.5	5.6	78.1	10.5	17.1
Social visit	3.8	4.5	6.2	60.5	14.3
Leisure	1.2	0	3.1	5.2	51.4

Table 8. Prediction results (%).

locations, 10.1% are wrongly predicted as non-work obligatory or social visit activities if they are accessed infrequently during weekdays. All the corresponding users work/study at multiple places, and the misidentified locations are their additional work/school places. Another 10.1% are mistaken as homes, if they have high visit frequencies at weekend. For instance, one of these users has two labelled work locations. They were visited at rates of 32% during weekdays and 42% on Sunday, respectively. While the first one was correctly identified, the second one was wrongly predicted as home. This suggests that the work regime plays an important role in distinguishing work locations from homes. While most people work during weekdays, certain minorities work on different shifts, especially to weekends or nights, generating distinct activity and travel patterns from the main stream of the population. (3) Non-work obligatory. The activities have low visit frequencies and short duration. The misclassification of the activities can be partially attributed to a combination of heterogeneity within this category. The various detailed types of the activities are likely performed at spatially independent locations and temporally varied preferences. For instance, shopping is mostly done in later time of the day than service or bringing/picking up activities. (4) Social visit. The activities are profiled with a middle level of visit frequencies during weekdays. If the locations are accessed less, they tend to be annotated as leisure or non-work obligatory activities; if more, they are considered as home or work/school places. The limited predictability could be caused by the underlying structure of an individual's social network, in which various degrees of relationship exist, ranging from closed one they visit routinely to the one they just meet occasionally. This generates variations in spatial-temporal features of the locations. (5) Leisure. Leisure activities are conducted in various places and at different time for an individual; they exhibit the lowest level of regularities and thus are the most challengeable to annotate. Apart from the spatial-temporal irregularities, the examination into two falsely predicted leisure locations reveals additional causes for the misclassification. The first one has a visit frequency of 36.3% in both the afternoon and evening on weekdays. It was the second most visited place for the corresponding user who has accessed this place 170 times over 337 days, such that 1 in 2 days he/her was observed there. This location is originally labelled as a restaurant; however, the call records suggest a high probability that he/her may work there instead of eating as a customer. The second location was ranked as the most visited place for the concerned user. He/she has in total conducted 383 visits over 442 days during both weekdays and weekends as well as at night. Nearly three in 4 days, he/she made calls there. Furthermore, the user has five locations collected in the training dataset, but none of which is labelled as home. This location is documented as sports; however, for this particular user, it is likely that this place is a home rather than a recreation site. While further investigation into the above two typical cases is needed before any definite conclusions are drawn, they nevertheless illustrate that our annotation method based on underlying activity and travel behaviour can effectively predict the activities, which are tailored to each individual. A location may have a single or multiple functions, but people visiting there could have different purposes. The match with geographic information alone is not able to identify this distinction. We shall call the location annotation at the individual level as *micro-location-annotation.*

6. Conclusions and future research

In this study, a cell phone location annotation method has been developed based on spatial-temporal regularities as well as sequential information intrinsic to activity and travel behaviour. The method does not depend on additional sensors and geographic details. The data requirement is simple and its collection cost is low. It is also generic to be transferable to other areas. On top of that, the method is independent of precisely geometric positions of individuals, thus considerably reducing privacy concerns.

Experiments on the annotation method using data collected from natural phone communication of users have achieved 76.6% prediction accuracy. With this probability, the activity conducted at a location for a user can be predicted by the spatial-temporal features of the visits disclosed by his/her call records. Furthermore, this study also shows the added value of the integration between machine learning methods and underlying activity and travel behaviour when annotating the location traces.

Nevertheless, despite the spatial-temporal regularities, activity locations still share commonalities in these two dimensions at a certain degree. Activity and travel behaviour is not solely decided by spatial-temporal elements, it is also affected by socio-economic conditions. The first improvement in future research should thus take this general background information into account. In particular, to address the potential causes for misclassifications of home and work/school locations, the annotation should be combined with the information on the number of home and work/school places of users as well as their work sectors and regimes. A broad picture of users' social networks, obtained from direct surveys and/or social networking sites, would strengthen the prediction of social visit activities. For non-work obligatory and leisure activities, the detailed types in each of these two categories should be handled separately, if a sufficient size of training data for the detailed types is available. The second improvement lies in finding an effective way of annotating locations, which are visited for multiple purposes for a particular user. While this study links the most frequent activity to a location, it dismisses additional activity types, which are performed by the user at different parts but within a same cell. In the training dataset, 5% of all the locations are visited for multiple purposes.

Today when simple phones are still prevalent constituting nearly 85% of total global handsets in use, this research makes undoubtedly an important contribution to the semantic explanation

http://dx.doi.org/10.5772/intechopen.70255

of the movement data. With the development of smart phones, the data from additional sensors installed on the phones will provide a third possibility of improvement by integrating the contextual information into the annotation process.

Author details

Feng Liu[1]*, JianXun Cui[2], Davy Janssens[1], Geert Wets[1] and Mario Cools[3]

*Address all correspondence to: feng.liu@uhasselt.be

1 Transportation Research Institute (IMOB), Hasselt University, Diepenbeek, Belgium

2 School of Transportation Science, Engineering, Harbin Institute of Technology, Harbin, China

3 LEMA, University of Liège, Liège, Belgium

References

[1] Song CM, Koren T, Wang P, Barabási AL. Modeling the scaling properties of human mobility. Nature Physics. 2010;**6**:818-823

[2] Liu F, Janssens D, Cui JX, Wang YP, Wets G, Cools M. Building a validation measure for activity-based transportation models based on cell phone data. Expert Systems with Applications. 2014;**41**(14):6174-6189

[3] Pink O, Hummel BA. Statistical approach to map matching using road network geometry, topology and vehicular motion constraints. In: The 11th International IEEE Conference on Intelligent Transportation Systems; 12–15 October 2008; Beijing, China. IEEE; 2008

[4] Laurila JK, Gatica-Perez D, Aad I, Blom J, Bornet O, Do TMT, Dousse O, Eberle J, Miettinen M. The mobile data challenge: Big data for mobile computing research. In: Proceeding Mobile Data Challenge (by Nokia) Workshop; 18–19 June 2012; Newcastle, UK. 2012

[5] Sae-Tang A, Catasta M, McDowell LK, Aberer K. Semantic place prediction using mobile data. In: Mobile Data Challenge (by Nokia) Workshop; 18–19 June; Newcastle, UK. 2012

[6] Liu F, Janssens D, Cui JX, Wets G, Cools M. Characterizing activity sequences using profile hidden Markov models. Expert Systems with Applications. 2015;**42**(13):5705-5722

[7] Cui JX, Liu F, Janssens D, An S, Wets G, Cools M. Detecting urban road network accessibility problems using taxi GPS data. Journal of Transport Geography. 2016b;**51**:147-157

[8] Cui JX, Liu F, Hu J, Janssens D, Wets G, Cools M. Identifying mismatch between urban travel demand and transport network services using GPS data: A case study in the fast growing Chinese city of Harbin. Neurocomputing. 2016a;**181**:4-18

[9] Kohavi R, John G. Wrappers for feature subset selection. Artificial Intelligence. 1997;**97**(1-2): 273-324

[10] Hall MA. Correlation-Based Feature Subset Selection for Machine Learning. New Zealand: Hamilton; 1998

[11] Quinlan JR. C4.5: Programs for Machine Learning. San Mateo, California: Morgan Kaufmann; 1993

[12] Breiman L. Random forests. Machine Learning. 2001;**45**(1):5-32

[13] Le Cessie S, van Houwelingen JC. Ridge estimators in logistic regression. Applied Statistics. 1992;**41**(1):191-201

[14] Keerthi SS, Shevade SK, Bhattacharyya C, Murthy KRK. Improvements to Platt's SMO algorithm for SVM classifier design. Neural Computation. 2001;**13**(3):637-649

[15] Liao SH, Chu PH, Hsiao PY. Data mining techniques and applications – A decade review from 2000 to 2011. Expert Systems with Applications. 2012;**39**(12):11303-11311

[16] Witten IH, Frank E, Hall MA. Data Mining: Practical Machine Learning Tools and Techniques. Burlington: Elsevier Inc.; 2011

Chapter 6

Examining Merchants' Refusal to Adopt Mobile Payment Systems in Spain

Francisco Liébana-Cabanillas,
Francisco Muñoz Leiva and Juan Sánchez Fernández

Additional information is available at the end of the chapter

http://dx.doi.org/10.5772/intechopen.70284

Abstract

Over the past years, traditional company management has been undergoing major changes regarding the adoption, implementation, and development of new technologies. Even if Internet commerce has the potential to revolutionize consumer behavior and the way merchants communicate with their customers, it is true that several activities related to the new technologies are still in the early stages of development or implementation. The main purpose of this study is to assess the determinants of m-payments from the point of view of merchants through an exploratory and qualitative analysis (literature review, focus groups, and in-depth interviews) in order to find the drivers and deterrents influencing the use of mobile payment systems in retail business. In order to properly approach the proposed research, a theoretical review of the actual situation of the different mobile payment systems across the different markets was carried through several personal interviews with merchants in the first place and, secondly, surveying over 151 retail companies in Spain. Conclusions and implications are discussed from the data and results drawn from this research, suggesting strategies to overcome some of the identified barriers and deterrents while also proposing some suggestions for future research opportunities.

Keywords: mobile payment, m-payment, merchant, adoption, barriers

1. Introduction

Development, adoption, implementation, and acceptance of the latest technologies are critical factors which have largely influenced business administration and management. Companies regard electronic commerce, or simply e-commerce, as the tool with the highest potential nowadays to revolutionize customers' purchasing habits and patterns while also impacting the different communication channels for merchants and their customers. However, some other business activities related to e-commerce are still in an early implementation stage or already maturing, the latter being the case of mobile business and mobile payment [64].

In recent years, mobile devices such as smartphones, personal digital assistants (PDAs), tablets, and laptops have been increasingly approached as means to transmit and receive all kinds of different data. In this sense, these devices have also seen an increase in their use as tools to facilitate the payment of different goods and services using their data transmission and reception capabilities; these payment systems are known as *mobile payments* or, simply, *m-payments*. m-Payment can thus be described as the completion of a financial transaction or purchase between individuals or other entities using a fast, convenient, easy, and secure tool anywhere from a mobile device. These new payment systems enjoy some proven advantages for both merchants and customers over other alternative payment systems such as the electronic point of sales (EPOS). Some of the aforementioned advantages would be (1) high level of adaptability and flexibility when taking into account the large amount of mobile phones and other mobile devices in the market, (2) fast transactions, (3) higher level of comfort and convenience resulting in a time-saving process, (4) possible classification and profiling of the different customers enabling custom strategies for the sale of goods and services, (5) lower cost of operations with reduced discount rates, etc. In the case of customers, the best regarded advantages of these payment systems are (a) the higher level of security of the involved economic transactions thanks to the use of technologies such as Global System for Mobiles (GSM), Universal Mobile Telecom System (UMTS), and, also, the Subscriber's Identity Module (SIM) card of the mobile device allowing for an improved encryption of data transmitted and received during the different transactions; (b) improved reliability of the payment system; (c) improved availability and offering of the goods and services which can be paid with these new tools; (d) reduced queuing and waiting times at the points of sale; and (e) lower rate of occurrence of incorrect transactions [39].

Traditional, single-channel business activities have been continuously evolving into multichannel operations in multiple markets across the world [72], with new retail formats facilitating the interactions between merchants and customers while increasing revenues for the respecting business companies involved [53]. In this sense, the use of smartphones has blurred and even started to eliminate the differences between the online and offline interactions between customers and merchants. The so-called Information Society is seeing a rapidly increasing growth every year; according to a recent report issued by Fundación Telefónica España (2015), mobile telephony reached a level of penetration of 95.5 subscriptions per 100 inhabitants worldwide in 2014 (2.4% raise over 2013), establishing an astonishing figure of over 6.6 billion for the total number of mobile telephony subscribers in the world. In this regard, Europe is the continent with more mobile subscribers (120 subscriptions per 100 inhabitants). Respecting m-payments, according to a report regarding e-commerce issued by the Online Business School [56], 27% of the total of online purchases in 2013 had been completed using one of the mentioned devices; this means a 55% raise over 2013 in the Spanish market. On the other hand, a recent research proposed by PayPal and carried out by an independent marketing research company (IPSOS) [33] concluded that mobile phone commerce saw an estimated growth of around 48% in Spain during that year, a raise rate well above the rate forecasted respecting the growth of online transactions in Spain during the same period of time. This report also found the best regarded

http://dx.doi.org/10.5772/intechopen.70284

features and characteristics that customers appreciate while purchasing goods and services through "smartphones" and "tablets": fast payments (36% of users mentioned this feature as one of the most relevant advantages), not needing to carry a physical wallet (24%),simplification of the payment process (22%), innovativeness nature of the payment system (21%), immediate confirmation of valid payments (20%), easy to use (19%), and, finally, not needing to share personal financial data with the different merchants (16%).

On a side note, the use of mobile payments is also seeing a considerable growth as reported by Capgemini [7] in their World Retail Banking Report analysis. In this sense, it is worth noting that this report found "only" 1.3 billion credit and debit active accounts as opposed to the remarkable statistics identifying well over 5 billion active mobile telephony subscriptions. According to Omlis (a provider of mobile payment solutions), this is an ideal, relevant scenario for a potential and major application of mobile payments. Juniper Research predicted that the number of users of mobile payment systems in 2013 (245 million) would duplicate by 2017 seeing a total number of over 450 million of users of mobile payments. Respecting the value of the *m-commerce* market, the market research company Gartner estimated a total amount of $507 billion accountable for mobile phone transactions in 2014. Therefore, all these studies show that the worldwide rate of adoption of mobile payments is also increasing rapidly, with some specific mediators influencing this increase such as the different tools available to consumers in order to access the new technologies, changes in lifestyle trends, and other economic factors.

From the perspective of the merchants and their offerings, according to a report issued by Tecnocom [70] on the different trends regarding payment systems, this report affirms that after assessing the actual demand of users respecting the different electronic payment systems in the case of the Spanish market, the use of mobile phone payments did not manage to establish itself as a particular strong alternative to the use of other systems. The information collected through the aforementioned reports and research allows observing certain differences between the needs of the users/customers regarding their purchases and the speed at which the market is adapting to said needs. In light of all of these findings, the main purpose of this research is to assess and evaluate the different factors influencing mobile payment from the perspective of the merchants through an exploratory, qualitative, and quantitative analysis (including a comprehensive literature review and approaching, focus groups, and in-depth interviews) aimed to identify the drivers and barriers to the use of the different mobile payment tools at the points of sale. This study also incorporates a section discussing conclusions and implications drawn from the results in order to overcome some of the identified deterrents while also proposing suggestions for future research opportunities.

2. Adoption of mobile payments: drivers and barriers

Mobile payment is considered by many experts as one of the "killer" or "star" applications with the greatest potential in the mobile communication sector ([26, 32]; and [55]). In this sense, mobile payment can be defined as any type of individual or business activity involving an electronic device capable of connecting to a mobile network in order to successfully complete an economic transaction [39].

Barriers to the adoption of mobile payments	Users' deficient knowledge of the new technology	No actual, real demand for this kind of payment system	Perceived trust/risk in the new payment system	Cost of adopting the payment system	Perceived lack of security	Technological issues and struggles
Chellappa and Pavlou [11]			x			
Claessens et al. [13]			x			
Begonha et al. [6]		x		x		
Pousttchi et al. [58]			x			
Siau and Shen [65]			x			x
Herzberg [31]			x			
Frolick and Chen [21]		x				
Wang and Cheung [74]		x				
Gebauer and Shaw [23]		x				
Misra and Wickamasinghe [51]			x			
Teo et al. [71]		x	x	x		
Mallat and Tuunainen [46]		x				x
Dewan and Chen [18]			x		x	
Liu et al. [44]			x			
Agarwal et al. [1]			x			
James and Versteeg [35]	x					
Chen [12]			x			
Balan et al. [3]						x
Islam et al. [34]			x			
Masamila et al. (2010)			x			
Wu et al. (2010)		x	x			
Islam et al. [34]	x					x
Saidi [63]	x					
Becher et al. [5]			x			
Little [43]	x					
Andreev et al. [2]			x			
Chang [9]			x			x
Chang [10]			x		x	
Slade et al. [68]			x			
Ramakrishna and Naik [59]			x			x

Barriers to the adoption of mobile payments	Users' deficient knowledge of the new technology	No actual, real demand for this kind of payment system	Perceived trust/risk in the new payment system	Cost of adopting the payment system	Perceived lack of security	Technological issues and struggles
Xin et al. [75]			x			
Slade et al. [67]			x			
Liébana-Cabanillas et al. [41]			x			
Ramos-de-Luna et al. [61]					x	
Ndege [54]			x	x		
Liébana-Cabanillas et al. [40]			x			
Ramos-de-Luna et al. [60]					x	
Komo et al. [37]				x	x	x
Liébana-Cabanillas et al. [42]			x			
Total	**4**	**8**	**26**	**5**	**5**	**8**
% total	**7.14**	**14.29**	**46.43**	**8.93**	**8.93**	**14.29**

Table 1. Barriers to the adoption of mobile payments.

As previously discussed, mobile payments constitute a recent innovation in its early stages of development and growth [77], yet they are widely extended in our society [28]. This marked presence generates an interest in finding the benefits and drawbacks for users of this new technology. With that purpose, this research reviewed the extant literature assessing different studies on this particular subject. After reviewing the scientific literature in this regard, this study found that the main barriers and deterrents to the adoption of mobile payments are the lack of trust in the new technology and the perceived risk when approaching it (see **Table 1**); 46.43% of users mentioned these factors during the survey process. Other relevant barriers that this study found are the lack of an actual demand for this type of services and the possible technological issues derived from their use (mentioned by 14.29% of respondents). Lastly, other variables such as the cost of adopting the new payment systems (8.93%), the perceived lack of security (also 8.93%), and the scarce knowledge users have regarding this new technologies (7.14%) are also mentioned as factors impacting to a lesser extent the intention to use. On the other hand, the main drivers for the adoption of mobile payments (see **Table 2**) are the convenience, comfort, and familiarity perceived by users when approaching the new technology through their smartphones (mentioned by 18.18% of respondents). Other significant drivers are attributes and characteristics such as ubiquity, personal nature, security, and the high penetration rate (13.64% across all these factors). Finally, this research also found less relevant drivers in variables such as mobility (9.09%) and compatibility (4.55%) [45].

Drivers for the adoption of mobile payments	Ubiquity	Personal nature	Mobility	Perceived security	Increased business operations and higher income	High penetration rate	Familiarity, convenience, and comfort	Compatibility
Clarke [14]	x							
Begonha et al. [6]		x			x			
Kreyer et al. [38]					x			
Frolick and Chen [21]	x		x					
Dourish [19]						x		
Mallat and Tuunainen [46]	x		x					
Teo et al. [71]		x						
Jarvenpaa and Lang [36]		x						
Heim and Sinha [30]						x		
Sahut [62]				x				
Meyer [50]						x		
Mallat et al. [47]								x
Chang [9]				x			x	
Ramakrishna and Naik [59]				x			x	
Chang [10]					x		x	
Tavilla [69]				x			x	
Total	**3**	**3**	**2**	**3**	**3**	**3**	**4**	**1**
% total	**13.64**	**13.64**	**9.09**	**13.64**	**13.64**	**13.64**	**18.18**	**4.55**

Table 2. Drivers for the adoption of mobile payments.

3. Methodology

The purpose of this study is to empirically evaluate the different factors influencing merchant adoption of mobile payment systems. In order to fulfill this purpose, our research approaches a sequential quantitative and qualitative analysis based on three different stages.

In the first stage of our research (carried out in the second fortnight of May in 2015), our research analyzed different databases of related scientific publications in order to evaluate an overview of the use of mobile payment systems in the market for daily commercial activities. Through a qualitative analysis based on two focus groups were established after this initial process; one of the groups involved the managers of payment systems of five financial entities in Spain, while the other focus group incorporated managers of five different commercial establishments). Both groups were surveyed with a concise, clear questionnaire designed specifically for this research.

After the successful completion of the first stage, our study focused on the second stage which was carried out roughly around the same time. In this stage, our research assessed the situation regarding the new payment systems in 25 different commercial establishments through another qualitative analysis in order to examine the actual reliability of the questions which were designed in the first stage. After successful completion of the first and second stages, we modified the questionnaire employed in our research in order to incorporate more significant questions respecting the following variables: knowledge regarding the new payment systems, different types of mobile payments, the use of each of the different mobile payment technologies, main providers of mobile payment tools, perceived usefulness of the different mobile payment systems, drivers and barriers to the use of mobile payments, and, lastly, the intention to use of these new payment solutions.

In the final stage of our research, carried out in June and July in 2015, we performed a qualitative analysis after identifying and validating all related factors to our study in previous publications and checking every fit commercial establishment for the purpose of this research. In this third stage, we performed the personal interviews in the commercial establishments employing a questionnaire which would empirically assess other questions relevant to this research.

A total of 400 different merchants were initially identified as fit for the purpose of this research. These merchants were subsequently classified according to their business activity and their contribution to the GDP of Spain. It is worth mentioning that once we contacted all of these 400 establishments, only 151 decided to contribute to our full (qualitative and quantitative) research (37.75% of total merchants approached).

We performed a batch of semi-structured interviews with an average completion time of around 50–75 min; these interviews were then transcribed and coded for later use. After the personal interview, participants were surveyed through an additional questionnaire which aimed to further improve data collection in order to successfully achieve the purpose of the questions used during the interviews.

The different profiles of the merchants participating in our research can be found in **Table 3**. The vast majority of these companies were identified as microenterprises since they had a low number of employees (between 1 and 9). This assessment is consistent with the data gathered by the Spanish Statistical Office, which reports the presence of a high number of companies in Spain as opposed to the average figures in other countries in the European Union. The difference is the smaller size of these Spanish companies than those in other countries in the EU (i.e., 76.8% smaller in terms of the number of employees and with an average income 72.8% lower). Also, according to the same data from the Spanish Statistical Office, the most relevant sector contributing to the GDP of Spain would be the traditional sector (47.4% of the total contribution) followed by the restoration sector (25%). These findings identify the service sector as the main performer regarding Spanish GDP.

	Categories	Frequency	Percentage/interval
Sector	Digital means	4	2.6
	ICTs (computers, telecommunications, software, etc.)	9	5.9
	Traditional (newspapers, cinema, etc.)	4	2.6
	Retailers	72	47.4
	Mail order or sales on the Internet	3	2.0
	Restoration	21	13.9
	Others	38	25.0
Company employees	0–9	116	76.8
	10–49	13	8.6
	50–249	7	4.6
	250–499	5	3.3
	500 or more	10	6.6
Company income in 2014	Under 2 million Euro (microenterprises)	110	72.8
	Between 3 and 10 million Euro (small enterprises)	8	5.3
	Between 11 and 50 million Euro (medium enterprises)	5	3.3
	Over 50 million Euro (large companies)	7	4.6
	Unknown	21	13.9
Sales channel employed	Physical store	141	93.4
	Internet	4	2.6
	Other (mail order or direct sale)	6	4.0
Position of the interviewee in the company	Company owner	49	32.5
	Company senior management	6	4.0
	Company middle management	4	2.6
	Store manager	12	7.9
	Store expert	6	4.0
	Employee	74	49.0
Experience	Average years with traditional payment systems	12.8	0–35
	Average years with mobile payment systems	0.16	0–4

Table 3. Respondent companies.

Respecting the different sale channels approached by the merchants contributing to our research, it is worth noting that the traditional channel is still the channel of choice employed by the majority of the companies participating in this study, with a sizable advantage over

http://dx.doi.org/10.5772/intechopen.70284

the different alternative channels. Our research also found that the average experience of respondents with traditional payment systems was an average of roughly 12.8 years. On the other hand, their average experience with mobile payment systems was significantly lower than that, not even amounting for a full year. This finding corroborates the relevance of our research when assessing the different determinant factors that might drive the acceptance and adoption of the new payment systems.

In order to complete the main questionnaire, other questions were incorporated regarding the level of knowledge of these new payment systems, their perceived trust and utility, and also some additional questions relevant to the nature of the different providers of mobile payment services.

4. Discussion and assessment of results

In order to properly assess results drawn from this research, a semantic analysis of the different questions used for the personal interviews was approached along with statistical quantitative analysis through SPSS v22 software.

4.1. Assessment of knowledge regarding mobile payment systems

The level of knowledge of the different mobile payment systems was found to be average (roughly 50%) after assessing the different respondent merchants. We found a variety of responses when participants were questioned about their knowledge respecting the different mobile payment systems; below some of the answers are included:

- *I know about mobile payment systems, but I do not use them.*
- *No, unfortunately I do not really know about mobile payment systems.*
- *I know they accept payments; I have seen a delivery person (of a shipping company) using them.*
- *I do not really know about mobile payment maybe that my smartphone can be used as a payment tool the same way as a credit card works; in our establishment we have a wireless point of sale.*
- *Yes, they are mostly contactless payment tools using apps installed in NFC-enabled smartphones.*
- *My knowledge is rather poor; I know a smartphone is required, but I have never used that payments.*
- *I only know mobile phone payment tools are some other ways to complete payments.*
- *I believe they refer to the services facilitating payments and collections using some apps for smartphones.*
- *Smartphones can be used in a similar way as credit cards.*
- *These payments allow using an app for smartphones to complete a payment.*
- *They are payments through a device connected to a network, especially smartphones. These payments are linked to credit accounts.*
- *I heard something on the news; they are new payment tools; we just need to place our smartphones in close proximity to the payment device in the establishment to complete a transaction.*

- *To be honest, I have never heard about mobile payment systems or how they would operate.*
- *Yes, mobile payment systems are employed in mobile devices which have been previously associated with a banking account; this is how payments are finally completed.*
- *They are a payment system using tools provided by the different banks, mostly a certain sticker you can attach to a mobile phone in order to complete payments involving amounts lower than 20 Euros ($21 approx.) In this case, there is no need to use a PIN code or a digital signature. If the amount is higher than 20 Euros, then a validation is required by placing the device in close proximity to the payment terminal.*
- *I consider mobile payment systems as modern and convenient, but I still believe they are difficult to implement successfully even if they enable a fast tool for customers to pay comfortably.*
- *I only know mobile payments use some kind of contactless, scanning device on the data-phone terminal in the establishment.*

As the answers above show, we can classify merchants into two groups: those with the appropriate knowledge of mobile payment systems and have already adopted them and, on the other hand, merchants which are basically oblivious regarding these new payment tools. In light of these findings, especially with so many merchants ignoring mobile payment systems, providers of mobile payment services should encourage the development of information campaigns and strategies in order to drive the adoption and management of these payment services.

4.2. Typology of the different mobile payment systems

Respecting the different payment systems accepted in the establishments which this research assessed and interviewed, only 14.57% would confirm the adoption of a certain mobile payment system, specifically NFC payment systems (see **Figure 1**).

In light of these results, and as we stated above, providers should encourage information strategies which will positively influence the adoption of mobile payment systems.

4.3. Mobile payment providers

Regarding the providers of the payment systems available to customers of the different merchants participating in this research, the vast majority of the respondents favored payment systems provided and supported by their associated financial entities (60.93%). On the other hand, the rest of respondents were undecided. Some of them adopted payment systems provided by external technology providers (14.57%), whereas the remaining respondents were indifferent to this matter (24.50%). In light of these findings, financial entities can be considered as the main providers of the payment systems adopted by the merchants. Therefore, these entities are expected to invest to a great extent in the different information campaigns regarding the new payment tools (**Figure 2**).

4.4. Perceived usefulness of mobile payment systems

Regarding the perceived usefulness of these new payment systems, this research assessed the answers provided by the different merchants with the following results: 64.90% of respondents did manifest a high perceived usefulness. On the other hand, one-quarter of the respondents (26.49%) did claim that the usefulness of these payment systems was minor and insignificant (**Figure 3**).

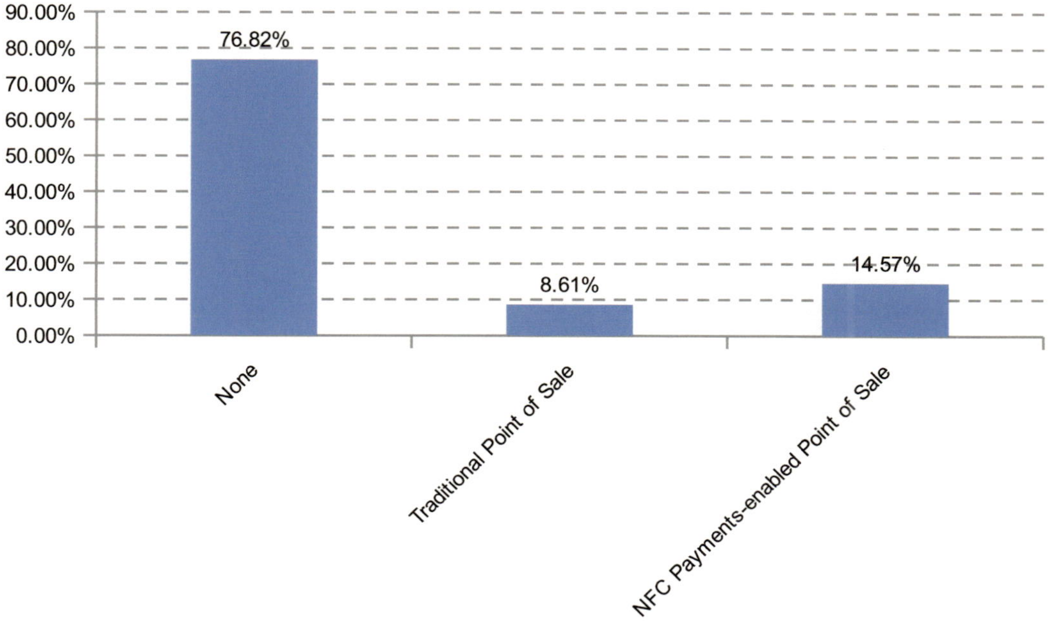

Figure 1. Typology of the different payment systems accepted by participating merchants.

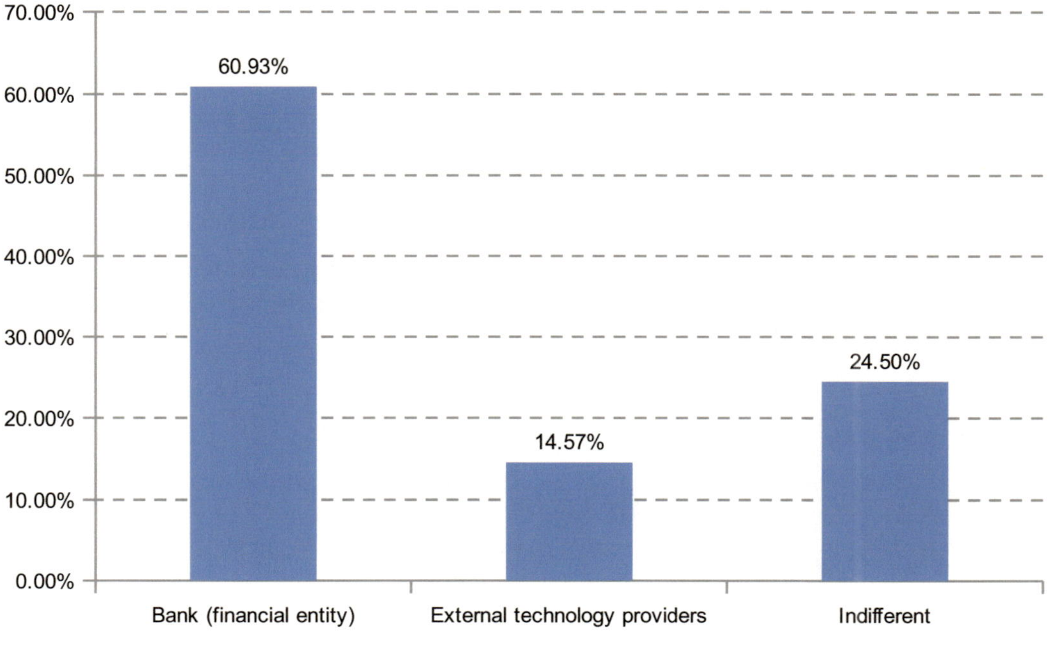

Figure 2. Preferred providers of mobile payment systems.

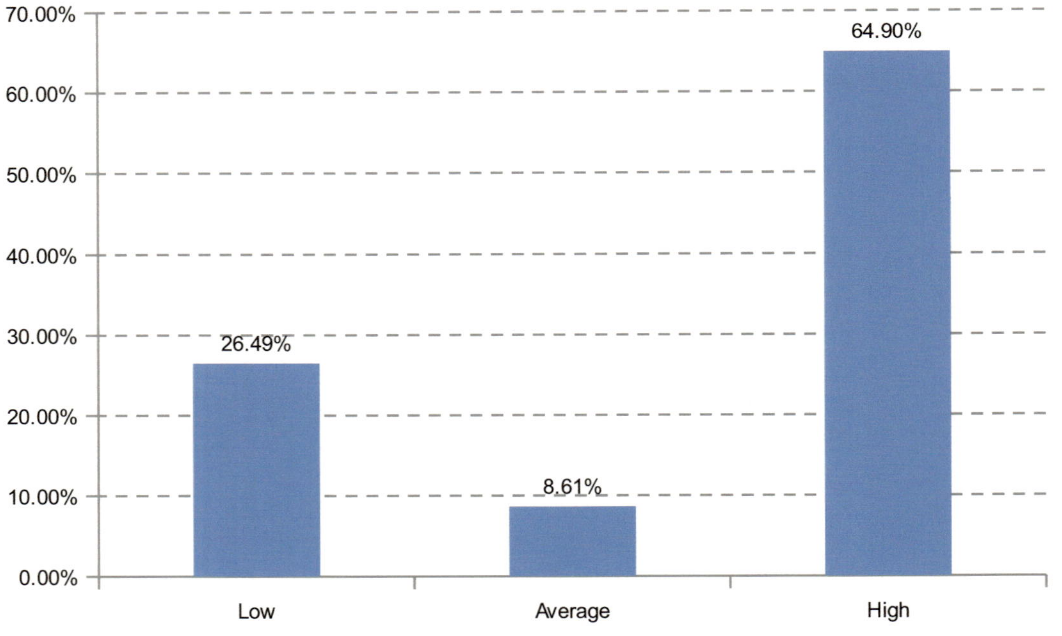

Figure 3. Perceived usefulness of the different mobile payment systems.

Relevant feedbacks from participating merchants were:

- *Mobile payments are rather useful allowing for instant payments anywhere.*
- *These new tools will be useful for me since I will be able to collect payments anywhere.*
- *Customers will benefit the most from the new payment systems; they will pay comfortably.*
- *This is a new trend and we have to adapt accordingly.*
- *These payments mean that cash is no longer necessary and transactions will be faster.*
- *They will facilitate the procedures and paperwork involved.*
- *They ensure a greater convenience for customers while the business transactions are clearer.*
- *Everyone has some kind of mobile phone; with these new payment systems, our customers have a permanent, convenient access to their money; maybe this will prevent the additional charges we pay for the different transactions.*
- *They might be useful in the future.*
- *I believe they could be rather useful and convenient.*
- *These new tools would allow a faster, easier payment process for our customers, but those willing to use them should be properly informed in the first place.*
- *These payment systems would be rather convenient for our customers.*

- *The usefulness of these payment systems depends on the age of the average customer. Regarding our elderly clientele, these tools would not be that useful.*
- *Additional payment tools always benefit customers, but our business would also benefit from the improved availability of different payment systems.*

In spite of the high perceived usefulness of these payment systems, this research proposes that those financial entities involved in these new payment systems should reinforce their communication and information strategies following the findings exposed earlier in this study.

4.5. Drivers and barriers

After reviewing the extant literature and assessing our respondents (**Figures 4** and **5**), we found that the most relevant factors negatively influencing the intention to use are the customer's lack of knowledge regarding the new technology (mentioned by 33.1% of respondents), followed by the scarce demand of a proper information on the new tools (18.5%), the level of perceived trust (10.6%), the cost of adoption (9.3%), the perceived lack of security (6%),and other technological issues (2.6%). It is worth mentioning that 11.9% of the respondents could not identify a single barrier or deterrent to the adoption of these new payment systems.

On the other hand, respecting the factors driving the intention to the use of mobile payment systems (**Figure 4**), it is worth noting factors such as the convenience and speed associated with these new tools (58.9%), merchants' perceived increase in security regarding the use of the new technology (14.6%), and improved sales turnover (9.3%), among other factors (4%).

4.6. Benefits derived from the use of mobile payment systems

In regard of merchants' perceived benefits when adopting the new payment tools, this research detected two different profiles clearly distinguished from each other; 29% of respondents could not identify any benefit whatsoever when using the new payment systems in a real-world environment, whereas, on the other hand, the remaining 71% of respondents could indeed perceive significant advantages and benefits such as the following: convenience, speed, improved time management, and lower cost of operation. We especially remark these factors since they are precisely the drivers, which the reviewed scientific literature suggests as the most relevant in this field. In order to improve the adoption of these new payment systems, this research identified again the need of proper information campaigns especially designed for the different merchants in order to divulge the benefits and advantages associated with the use of the new tools.

4.7. Intention to use mobile payment systems

Finally, after assessing respondents' intention to use new mobile payment systems, this research found a patent intention to use them in only 17.88% of the total respondents, reflecting the incipient penetration of this new technology regarding the majority of merchants in Spain. On the other hand, 82.12% of respondents did show no interest whatsoever in adopting the new payment systems.

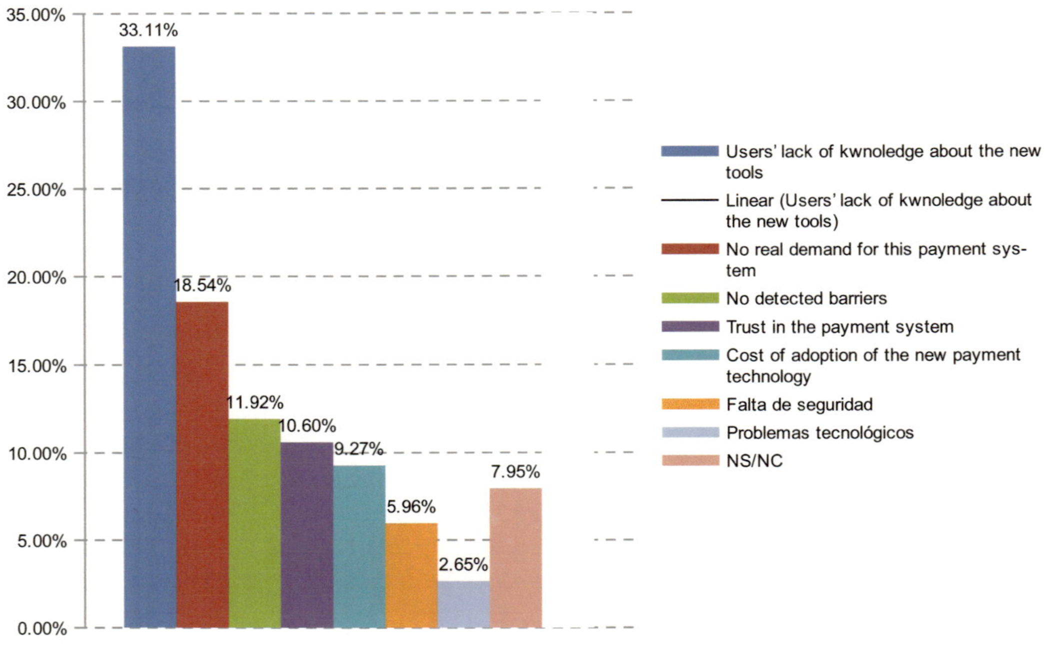

Figure 4. Barriers to the adoption of mobile payments.

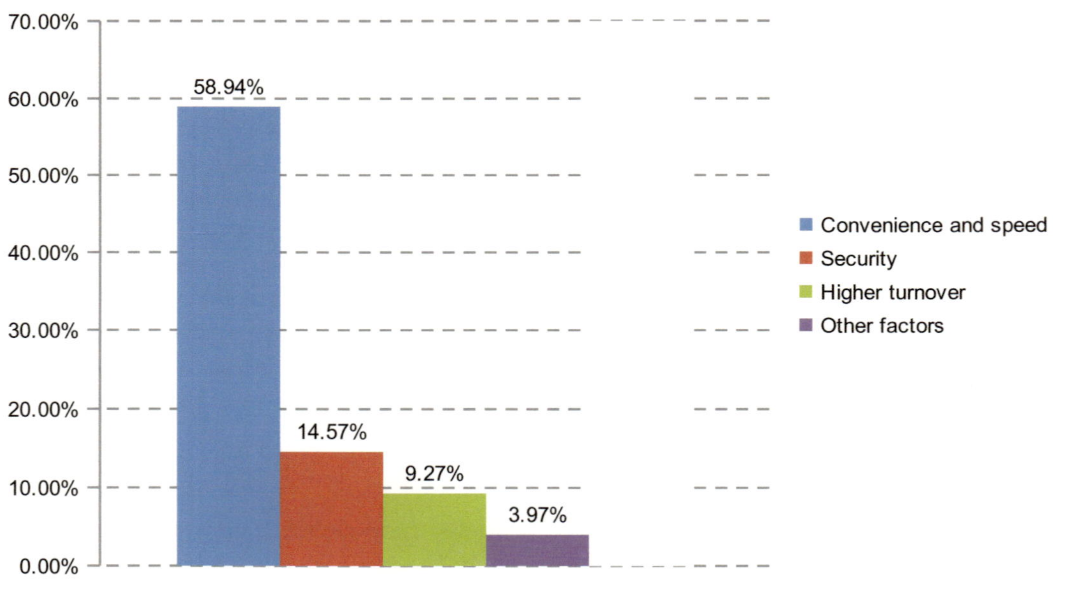

Figure 5. Drivers influencing the adoption of mobile payments.

5. The importance of trust and perceived risk

In the case of this research, both trust and perceived risk are significant drivers for merchant's adoption of mobile payment systems; thus they're thoroughly assessed.

In recent decades, research performed in the field of marketing has stressed the importance of trust among the different involved parties as a driver for the continuity and durability of a relationship. This is of great significance at a business level [22].

Trust was described by Singh and Sirdeshmukh [66] in the context of B2C e-commerce as "the psychological state leading to accept the vulnerability of a trustor, based on positive expectations of the trustee's actions." In this sense, Van Der Heijden et al. [73] reported that trust in the ambit of online purchasing could be defined as "the willingness of one of the parties (the purchaser) to be vulnerable to the actions of a virtual establishment, based on the expectations that this virtual establishment will carry out an important action for the customer or purchaser, regardless of his or her ability to conduct or control the virtual establishment."

Trust has traditionally involved two key dimensions: cognitive and behavioral. Dwyer et al. [20] approached the cognitive dimension of trust in order to describe it as "one party's expectation that the words or promises of the other party are reliable and that the other party will fulfill his or her obligations in a relational exchange." Regarding the cognitive approach, we detected in the extant literature three different moderating factors for beliefs: competence, integrity, and benevolence. These variables exhibit psychometric properties which are fit for the measurement scale approached [8]. On the other hand, Mayer et al. [48] and McKnight et al. [49] examined predictability (the capacity to predict someone else's behavior in any situation) as an additional moderating variable [52].

From a behavioral point of view, trust can be described as "the willingness of a trustor to be vulnerable to the actions of a trustee, based on the expectation that the trustee will perform a particular action important for the trustor, regardless of the capacity of the trustor to survey or control the trustee" [48], referring to the willingness to behave following a certain behavioral pattern. Multiple studies examine this variable in order to identify the acceptance success rate of new technologies [76] such as e-commerce.

Following a different approach, Bauer [4] carried out an analysis on perceived risk based on two different variables: Firstly, uncertainty, which can be defined as the lack of knowledge that consumers show when they are actually involved in the process of purchasing, and, secondly, the lack of information on the potential negative consequences of purchases. In this sense, this author also posited that consumer behavior is always connected to a certain risk; behavioral patterns involve effects that cannot be predicted with any certainty [4]. In this sense, Gupta and Kim [29] described perceived risk as "consumers' perception about the uncertainty and the adverse consequences of a transaction performed by a seller," while Gefen et al. [24] reported "the consequence of a decision reflecting the variation of its eventual results." Finally, the literature also includes the definition of perceived risk by Gerrard and Barton Cunningham [25], described as "the possibility that the use of an innovation could not be safe."

6. Managerial implications regarding the adoption of mobile payments

The use of payment tools has been related to our species since ancient times. In spite of this long established relationship, the adoption (involving implementation and use) of the new, modern payment systems in our current economic environment has greatly influenced the business scenario. Mobile phone payments can be considered as the most significant tool among the recent payment systems introduced in the market due to the relevance of wireless devices in our society, their simple accessibility, and the constant improvements in the technology associated with these new payment tools both for online contexts (mainly the Internet and social media) and traditional, "offline" applications such as swipe card readers, POS machines, etc.

In the past decades, the swift increase in the level of competitiveness and technological improvements involved in all commercial sectors, as factors which are becoming equally significant, has led to the development of a new communication strategy for companies and customers; this new channel evolved into a practical application known as e-commerce. In this sense, companies have been catering appropriately according to the respective demands of their business sector. Mobile commerce and the new mobile payment systems are relevant actors with a key role in the evolution of the market due to the widely extended use of mobile devices, especially smartphones, with the prospect of a high penetration rate in our society. For all these reasons, the business prospect for this particular sector is truly promising.

As observed from the data discussed earlier in this study, certain countries might not follow the general trends of the market; this appears to be the case regarding the market in Spain. Our research aims to identify the reasons for this situation from the perspective of the merchants. Research assessing the adoption and acceptance of the mobile payment systems is rather scarce and might improve future conceptions and prospects for the market of these new payment tools. According to the results obtained in this empirical analysis, we have outlined and proposed an appropriate research framework for merchant adoption of mobile payments. This is depicted in **Figure 6**.

In our opinion, on the basis of the results obtained from this research, the lack of information on the new payment tools significantly hinders their adoption. The development of proper information campaigns in this regard would help to overcome the barriers while also reinforcing the drivers affecting the final intention to use of mobile payment systems. This information should approach each and every one of the elements in the framework of mobile payment. In this sense, Dahlberg et al. [16] proposed that the relevance of mobile payments should be analyzed through Porter's Extended Rivalry Model [57] and the Generic Contingency Theory. With this in mind, we consider that the sources of this critical information should involve the

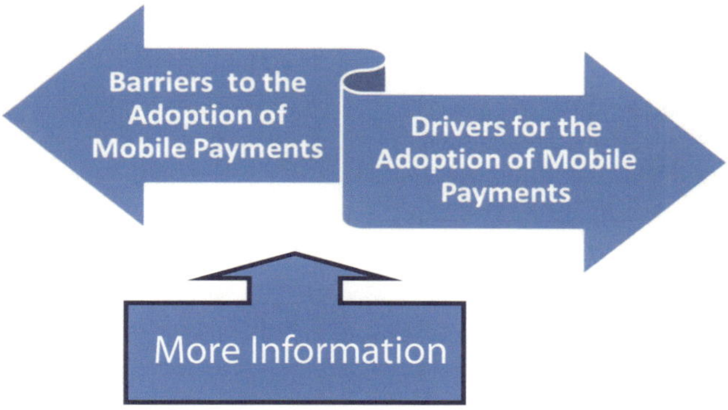

Figure 6. Managerial implications derived from the adoption of mobile payments.

main providers of mobile payment system and also, to a great extent, merchants interested in the new technology.

As for the companies providing mobile payment services (banks, financial entities, and external technology providers), they need:

a. To properly inform merchants through online (websites and social media) and traditional communication channels on the different mobile payment tools.

b. To offer joint promotions where the cost of the transactions would be partially borne by the financial entity or provider of the mobile payment tool, for instance, offering flat-rate plans for a certain period of time after the initial adoption of the new technology.

c. To advertise the benefits and advantages derived from the use of mobile payment systems.

d. To enforce and develop robust security systems for all consumer transactions made through these electronic payments in order to reinforce users' perceived trust.

e. To provide merchants new and modern point-of-sale terminals in order to speed up and facilitate in-store purchases.

f. To reinforce security systems in commercial establishments in order to increase merchants' perceived trust in the new technology.

g. To include launch promotions to facilitate the adoption of mobile payment systems by both merchants and final users.

As for merchants interested in the adoption of mobile payment systems, they need:

a. To properly inform customers on the different existing mobile payment tools through online (websites and social media) and traditional communication channels.

b. To advertise the benefits and advantages derived from the use of the new mobile payment tools.

c. To offer in-store promotions in order to encourage the use of mobile payment tools.

7. Final discussion, limitations, and future research opportunities

7.1. Discussion of results

Recent studies in the literature have thoroughly analyzed users' behavior in the adoption of the new mobile payment systems [27, 42, 67]. These studies have found and assessed the most significant barriers and drivers influencing the adoption of mobile payments. However, the focus of these research efforts was placed on consumers with no evaluation whatsoever from the perspective of the merchants. In addition, it is worth mentioning that most mobile payment adoption initiatives have failed before reaching consumers and merchants [15].

Even if a significant number of consulting firms are elaborating and delivering key, successful forecast studies predicting the future and potential behavior of customers toward the new mobile payment systems, real-world results show that these new payment tools have not yet really taken off and are actually in a difficult, serious situation derived from the conflict of interests of the different involved actors and the respective payment contexts.

In light of these findings, the aim of our research is to analyze the most significant barriers and drivers regarding the adoption of the new payment systems from the perspective of the merchants. In order to achieve this goal, this research approaches a qualitative and quantitative analysis carried out after a comprehensive literature review with the purpose of finding and examining the aforementioned factors affecting the use of mobile payment tools. As mentioned earlier in our study, research in this specific field of knowledge is scarce even if it could lead for brighter prospect regarding the future adoption of mobile payments.

This research corroborates the idea that the new technology is rather appealing for all actors involved in the respective market [17]. However, in terms of the benefits and advantages for the merchants, we believe that the new mobile payment tools would speed up the actual purchases while improving the security of the transactions at the same time, optimizing sales turnover and enabling new marketing strategies aimed toward smartphones. However, before reaching this point in the development and adoption of mobile payment systems, merchants need to overcome the different barriers and deterrents detected in this research.

This research has proven the importance of a certain set of deterrents and drivers affecting merchant adoption of mobile payment systems. Regarding the different barriers, the intention to use will definitely improve as long as users' knowledge of the new technology is also reinforced. For mobile payments to actually succeed in the current situation of the market, the following goals need to be achieved: (a) a real demand for this type of payments, (b) a higher perceived trust regarding the operation of the new technology, (c) optimizing and sharing the costs derived from the adoption of these payment systems, and (d) an enhanced security infrastructure to overcome possible technological issues related to the use and nature of mobile wireless devices. Finally, this research also found that perfecting the convenience, speed, security, and merchants' sales turnover associated with the use of the new payment systems leads to a considerably higher intention to use them.

7.2. Limitations and future research opportunities

Despite its contributions, this research also shows some limitations which provide fruitful avenues for future research. In the first place, as a preliminary study, this research only approached a basic statistical analysis. In this regard, researchers are currently developing a theoretical model aiming to examine the adoption of mobile payments by merchants across different countries. This model will also be contrasted and tested using structural equation modeling. This process should provide detailed results useful for researchers and users that could be projected to other markets. In addition, the sample employed in this research consisted only of Spanish merchants; in this regard, half of the actual respondents were employees of the different companies with no decision-making powers related to the adoption of the new payment technologies. In this sense, future research should approach the individuals in charge of this type of operations. In addition, an analysis following a longitudinal approach should facilitate the evaluation of the actual adoption instead of focusing solely on the behavioral intention. This type of analysis would also detect changes in the identified drivers and barriers over the continuous use of the mobile payment tools.

Author details

Francisco Liébana-Cabanillas*, Francisco Muñoz Leiva and Juan Sánchez Fernández

*Address all correspondence to: franlieb@ugr.es

University of Granada, Granada, Spain

References

[1] Agarwal S, Khapra M, Menezes B, Uchat N. Security issues in mobile payment systems. Computer Society of India. 2007:142-152

[2] Andreev P, Pliskin N, Rafaeli S. Drivers and inhibitors of mobile-payment adoption by smartphone users. International Journal of E-Business Research (IJEBR). 2012;8(3):50-67

[3] Balan RK, Ramasubbu N, Prakobphol K, Christin N, Hong J mFerio: The design and evaluation of a peer-to-peer mobile payment system. In: Proceedings of the 7th International Conference on Mobile Systems, Applications, and Services ACM. 2009. pp. 291-304.

[4] Bauer RA. Consumer behavior as risk taking. In Hancock R, editor. Proceedings of 43rd Dynamic Marketing for a Changing World. Chicago, IL: American Marketing Association. 1960. pp. 389-398.

[5] Becher M, Freiling FC, Hoffmann J, Holz T, Uellenbeck S, Wolf C. Mobile security catching up? Revealing the nuts and bolts of the security of mobile devices. In 2011 IEEE Symposium on Security and Privacy (SP). IEEE. 2011, May. pp. 96-111.

[6] Begonha DB, Hoffman A, Melin P. M-payments; hang up, try again. Credit Card Management. 2002;**15**(10):40-44

[7] Capgemini (2015). World Retail Banking Report 2015. Available in https://www.worldretailbankingreport.com/

[8] Castañeda JA. El comportamiento del usuario de Internet: Análisis de los antecedentes y consecuencias de la fidelidad. [Tesis Doctoral]. Departamento de Comercialización e Investigación de Mercados: Universidad de Granada; 2005

[9] Chang TK. A secure cloud-based payment model for M-commerce. In 2013 42nd International Conference on Parallel Processing (ICPP). IEEE. 2013. pp. 1082-1086.

[10] Chang TK. A secure mobile bill payment (MOBILL) service for M-commerce. A secure operational model for mobile payments. The Scientific World Journal. 2014

[11] Chellappa RK, Pavlou PA. Perceived information security, financial liability and consumer trust in electronic commerce transactions. Logistics Information Management. 2002;**15**(5/6):358-368

[12] Chen L. A model of consumer acceptance of mobile payment. International Journal Mobile Communications. 2008;**6**(1):32-52

[13] Claessens J, Dem V, De Cock D, Preneel B, Vandewalle J. On the security of today's online electronic banking systems. Computers & Security. 2002;**21**(3):253-265

[14] Clarke I III. Emerging value propositions for m-commerce. Journal of Business Strategies. 2001;**18**(2):133

[15] Dahlberg T, Guo J, Ondrus J. A critical review of mobile payment research. Electronic Commerce Research and Applications. 2015;**14**(5):265-284

[16] Dahlberg T, Mallat N, Ondrus J, Zmijewska A. Past, present and future of mobile payments research: A literature review. Electronic Commerce Research and Applications. 2008;**7**(2):165-181

[17] Dennehy D, Sammon D. Trends in mobile payments research: A literature review. Journal of Innovation Management. 2015;**3**(1):49-61

[18] Dewan SG, Chen LD. mobile payment adoption in the US: A cross-industry, crossplatform solution. Journal of Information Privacy and Security. 2005;**1**(2):4-28

[19] Dourish P. What we talk about when we talk about context. Personal and Ubiquitous Computing. 2004;**8**(1):19-30

[20] Dwyer FR, Schurr PH, Oh S. Developing buyer-seller relationships. The Journal of Marketing. 1987:11-27

[21] Frolick MN, Chen L. Assessing M-commerce opportunities. Information Systems Management. 2004;**21**(2):53-61

[22] García N, Santos ML, Sanzo MJ, y Trespalacios JA. (2008). El papel del marketing interno como antecedente de la capacidad de innovación de la PYME. Efecto sobre los resultados empresariales. XXII Congreso anual AEDEM. Salamanca, 18, 19 y 20 de Junio de 2008.

[23] Gebauer J, Shaw MJ. Success factors and impacts of mobile business applications: Results from a mobile e-procurement study. International Journal of Electronic Commerce. 2004;**8**(3):19-41

[24] Gefen D, Rao V, & Tractinsky N. Conceptualization of trust, risk and their relationship in electronic commerce: The need for clarifications. In: Proceedings of the 36th Hawaii International Conference on IS. 2003.

[25] Gerrard P, Barton Cunningham J. The diffusion of internet banking among Singapore consumers. International Journal of Bank Marketing. 2003;**21**(1):16-28

[26] Ghezzi A, Renga F, Balocco R, Pescetto P. Mobile payment applications: Offer state of the art in the Italian market. Info. 2010;**12**(5):3-22

[27] Guillén A, Herrera LJ, Pomares H, Rojas I, Liébana-Cabanillas F. Decision support system to determine intention to use mobile payment systems on social networks: A methodological analysis. International Journal of Intelligent Systems. 2016;**31**(2):153-172

[28] Guo X, Zhao Y, Jin Y, Zhang N (2010). Theorizing a two-sided adoption model for mobile marketing platforms. ICIS 2010 Proceedings, Paper 128, 1-16.

[29] Gupta S, Kim HW. Value-driven Internet shopping: The mental accounting theory perspective. Psychology & Marketing. 2010;**27**(1):13-35

[30] Heim GR, Sinha KK. Service product configurations in electronic business-to-consumer operations: A taxonomic analysis of electronic food retailers. Journal of Service Research. 2005;**7**(4):360-376

[31] Herzberg A. Payments and banking with mobile personal devices. Communications of the ACM. 2003;**46**(5):53-58

[32] Hu X, Li W, Hu Q. Are mobile payment and banking the killer apps for mobile commerce? In Proceedings of the 41st Annual Hawaii International Conference on System Sciences. IEEE. 2008. pp. 84-84.

[33] IPSOS (2015). Los medios de pago, un paisaje en movimiento. Disponible en https://www.pwc.es/es/publicaciones/financiero-seguros/assets/medios-pago-paisaje-movimiento.pdf

[34] Islam MA, Ahmad TSB, Khan MA, Ali MH. Adoption of M-commerce services: The case of Bangladesh. WORLD. 2010;**2**(1):37-54

[35] James J, Versteeg M. Mobile phones in Africa: How much do we really know? Social Indicators Research. 2007;**84**(1):117-126

[36] Jarvenpaa SL, Lang KR. Managing the paradoxes of mobile technology. Information Systems Management. 2005;**22**(4):7-23

[37] Komo L, Kyando E, Ngare P. Determinants of consumers' adoption of mobile parking payment services in Kenya. Journal of Emerging Trends in Economics and Management Sciences (JETEMS). 2016;**7**(1):1-12

[38] Kreyer N, Pousttchi K, Turowski K. Standardized payment procedures as key enabling factor for mobile commerce. In International Conference on Electronic Commerce and Web Technologies. 2002. pp. 400-409. Berlin Heidelberg: Springer.

[39] Liébana-Cabanillas, F. (2012). The role of payment systems in new electronic environments [doctoral dissertation]. Marketing and Markets Research Department, University of Granada.

[40] Liébana-Cabanillas F, Herrera LJ, Guillén A. Variable selection for payment in social networks: Introducing the Hy-index. Computers in Human Behavior. 2016;**56**:45-55

[41] Liébana-Cabanillas F, Muñoz-Leiva F, Sánchez-Fernández J. Behavioural model of younger users in M-payment systems. Journal of Organizational Computing and Electronic Commerce. 2015 just-accepted

[42] Liébana-Cabanillas F, Muñoz-Leiva F, Sánchez-Fernández J. A global approach to the analysis of user behavior in mobile payment systems in the new electronic environment. Service Business. 2017:1-40

[43] Little AD. (2011). Global M-Payment Report 2011. Available in www.adlittle.com

[44] Liu Y, Cao X, Dang L. Mobile Payment in Enabling Technologies for Wireless E-Business. Berlin Heidelberg: Springer; 2006. p. 233-252

[45] Mallat N, Dahlberg T. Consumer and merchant adoption of mobile payment solutions. In: Managing Business in a Multi-channel World: Success Factors for e-Business. 2005. p. 32

[46] Mallat N, Tuunainen, VK. Merchant adoption of mobile payment systems In International Conference on Mobile Business, 2005. ICMB 2005. 2005. pp. 347-353. IEEE.

[47] Mallat N, Rossi M, Tuunainen VK, Öörni A. The impact of use context on mobile services acceptance: The case of mobile ticketing. Information Management. 2009;**46**(3):190-195

[48] Mayer RC, Davis JH, Schoorman FD. An integrative model of organizational trust. Academy of Management Review. 1995;**20**(3):709-734

[49] McKnight DH, Cummings LL, Chervany NL. Initial trust formation in new organizational relationships. Academy of Management Review. 1998;**23**(3):473-490

[50] Meyer A. Cell phone banking will bring a microfinance revolution. Tech for Change. 2007:1-2

[51] Misra SK, Wickamasinghe N. Security of a mobile transaction: A trust model. Electronic Commerce Research. 2004;**4**(4):359-372

[52] Muñoz F. La adopción de una innovación basada en la Web [Tesis Doctoral]. Departamento de Comercialización e Investigación de Mercados, Universidad de Granada; 2008

[53] National Retail Federation (2011). Mobile Retailing Blueprint. A Comprehensive Guide for Navigating the Mobile Landscape. Available in www.nrf.com/mobile

[54] Ndege MD. Modelling uptake of mobile payments by MSMES in Kenya [doctoral dissertation]. University of Nairobi; 2015

[55] Ondrus, J, Lyytinen, K, Pigneur, Y. Why mobile payments fail? Towards a dynamic and multi-perspective explanation. In 42nd Hawaii International Conference on System Sciences, HICSS'09. IEEE. 2009. pp. 1-10.

[56] Online Business School (2014). El Comercio Electrónico 2014. Available in http://www.obs-edu.com

[57] Porter M. Competitive Strategy. New York, NY: Free Press; 1998

[58] Pousttchi, K, Turowski, K, Weizmann, M (2003, April). Added value-based approach to analyze electronic commerce and mobile commerce business models. In Andrade RAE, Gómez JM, Rautenstrauch C, Rios RG, editors. International Conference of Management and Technology in the New Enterprise. La Habana. pp. 414-423.

[59] Ramakrishna PG, Naik MC. A Secure Mobile Bill Payment (MOBILL) Service for M-Commerce. International Journal of Science and Research (IJSR). 2014;**3**(9):2234-2237

[60] Ramos-de-Luna I, Montoro-Ríos F, Liébana-Cabanillas F. Determinants of the intention to use NFC technology as a payment system: An acceptance model approach. Information Systems and e-Business Management. 2016;**14**(2):293-314

[61] Ramos-de-Luna I, Montoro-Ríos F, Liébana-Cabanillas F. Determinants of the intention to use NFC technology as a payment system: An acceptance model approach. Information Systems and e-Business Management. 2015:1-22

[62] Sahut JM. Electronic wallet in danger. Journal of Internet Banking and Commerce. 2006;**11**(2)

[63] Saidi E. Towards a faultless mobile commerce implementation in Malawi. Journal of Internet Banking and Commerce. 2010;**15**(1):1-13

[64] Sharma A, Sheth JN. Web-based marketing: The coming revolution in marketing thought and strategy. Journal of Business Research. 2004;**57**(7):696-702

[65] Siau K, Shen Z. Building customer trust in mobile commerce. Communications of the ACM. 2003;**46**(4):91-94

[66] Singh J, Sirdeshmukh D. Agency and trust mechanisms in consumer satisfaction and loyalty judgments. Journal of the Academy of Marketing Science. 2000;**28**(1):150-167

[67] Slade EL, Dwivedi YK, Piercy NC, Williams MD. Modeling consumers' adoption intentions of remote mobile payments in the United Kingdom: Extending UTAUT with innovativeness, risk, and trust. Psychology & Marketing. 2015;**32**(8):860-873

[68] Slade EL, Williams MD, Dwivedi YK. Devising a research model to examine adoption of mobile payments: An extension of UTAUT2. The Marketing Review. 2014;**14**(3):310-335

[69] Tavilla E (2015). Transit Mobile Payments: Driving Consumer Experience and Adoption.

[70] Tecnocom (2014). Trends in Payment Instruments 2014. Available in http://www.tecnocom.es/documents/10181/27549/Tecnocom2014_ingles.pdf

[71] Teo E, Fraunholz B, Unnithan C (2005, January). Inhibitors and facilitators for mobile payment adoption in Australia: A preliminary study. In International Conference on Mobile Business, ICMB 2005. pp. 663-666. IEEE.

[72] Torrent-Sellens J, Castillo D, Gabaldón P, Ruiz E, Sainz J. Hacia la bancamulticanal. Madrid: ESIC; 2010

[73] Van der Heijden H, Verhagen T, Creemers M. Understanding online purchase intentions: Contributions from technology and trust perspectives. European Journal of Information Systems. 2003;**12**(1):41-48

[74] Wang S, Cheung W. E-business adoption by travel agencies: prime candidates for mobile e-business. International Journal of Electronic Commerce. 2004;**8**(3):43-63

[75] Xin H, Techatassanasoontorn AA, Tan FB. Antecedents of consumer trust in mobile payment adoption. Journal of Computer Information Systems. 2015;**55**(4):1-10

[76] Yang MH, Lin B, Chandlrees N, Chao HY. The effect of perceived ethical performance of shopping websites on consumer trust. Journal of Computer Information Systems. 2009;**50**(1):15-24

[77] Zhu Y. A new architecture for secure two-party mobile payment transactions [thesis]. Department of Mathematics and Computer Science, University of Lethbridge; 2010

Chapter 7

Smartphone: The Ultimate IoT and IoE Device

Mehdia Ajana El Khaddar and
Mohammed Boulmalf

Additional information is available at the end of the chapter

http://dx.doi.org/10.5772/intechopen.69734

Abstract

Internet of Things (IoT) and Internet of Everything (IoE) are emerging communication concepts that will interconnect a variety of devices (including smartphones, home appliances, sensors, and other network devices), people, data, and processes and allow them to communicate with each other seamlessly. These new concepts can be applied in many application domains such as healthcare, transportation, and supply chain management (SCM), to name a few, and allow users to get real-time information such as location-based services, disease management, and tracking. The smartphone-enabling technologies such as built-in sensors, Bluetooth, radio-frequency identification (RFID) tracking, and near-field communications (NFC) allow it to be an integral part of IoT and IoE world and the mostly used device in these environments. However, its use imposes severe security and privacy threats, because the smartphone usually contains and communicates sensitive private data. In this chapter, we provide a comprehensive survey on IoT and IoE technologies, their application domains, IoT structure and architecture, the use of smartphones in IoT and IoE, and the difference between IoT networks and mobile cellular networks. We also provide a concise overview of future opportunities and challenges in IoT and IoE environments and focus more on the security and privacy threats of using the smartphone in IoT and IoE networks with a suggestion of some countermeasures.

Keywords: smartphone, Internet of Things, Internet of Everything, ubiquitous, context, sensors, RFID, security, privacy

1. Introduction

The Internet of Things (IoT) is a network of intelligent devices ranging from home appliances to industrial equipment that can become connected to the Internet, monitor themselves, send contextual information such as pressure, location, and temperature, and communicate somehow, anytime, anywhere on the planet (e.g., a milk carton sending sensor and identification information to a radio-frequency identification (RFID) reader when the temperature is getting higher than a threshold or when the milk carton is moved to a hot place) [1]. IoT means "connecting anyone, anything, anytime, anyplace, any service and any network" [1]. The concept of IoT has been extended by Cisco to Internet of Everything (IoE) to include in addition to things (machine-to-machine (M2M)), people (technology-assisted people-to-people (P2P)) and processes (machine-to-people (M2P)) interactions [2]. Cisco [2] defines Internet of Everything (IoE) as "the intelligent connection of people, process, data and things" [2], englobing interactions and communications generated by users while using a variety of networked devices (e.g., if a person forgot if s/he left the oven on at home, s/he wouldn't have to run back home to check it as s/he could just use a specific application and do it remotely using her/his smartphone) [2]. The proliferation of mobile connectivity and the decreasing prices of sensors and processors are encouraging the rapid growth of the IoT and IoE. Smart devices, for example, smartphones, smartwatches, PDAs, phablets, and tablets, will be the primary interaction tools used by people in a connected environment including cars, homes, and workplaces. Gartner expects in [3] that "the number of connected things and devices to rise to 25 billion by 2020 while other more aggressive estimates put the figure at 50 billion" [3]. "For this to be realized we need to have devices that are not only smart but should be able to access the Internet without being connected to a physical local area network (LAN) or wireless fidelity (Wi-Fi) network, should have an independent power source (e.g. battery), and should have the ability to sense the physical environment and send context information seamlessly" [3]. In today's world of emerging technologies, this could be made a reality: RFID, Bluetooth, 3G, 4G, 5G, wireless sensor networks, etc., along with long-lasting batteries, all bundled in one inexpensive, small, light, and portable device, which is the smartphone.

Equipped with the aforementioned technologies stated above, the smartphone gathers context data about the user (e.g., geolocation, temperature, health conditions, etc.) and interacts seamlessly with various devices using different types of connections such as Bluetooth, near-field communications (NFC), Wi-Fi, etc. Therefore, the smartphone can be considered as the user's ultimate device for IoT and IoE interactions and control. Big data, mobility, and cloud services are the principal parts of IoE concept, and using the smartphones everywhere is helping the IoE movement forward. Many services can be done in real time using the cloud and smartphones, for example, we can use our smartphone to order items online quickly, use an application to see if a specific store has an item in stock, or even better check how big is the queue in this store, order an item, then let customer services know that you are on your way to pick it up.

In this chapter, we will give an overview about the Smartphones' enabling technologies for the Internet of Things (IoT) and Internet of Everything (IoE), such as RFID, NFC, optical tags and quick response codes, Bluetooth, etc. We will also discuss the different application areas of IoT and IoE through the use of smartphones interconnected to other devices and show how the smartphone behaves in a cloud environment using different offered services. Finally, we will state the future opportunities and challenges of IoT and IoE applications. Some of the opportunities that will be discussed include context and ubiquitous services. Challenges will target basically the areas of privacy and security.

2. Smartphones

A cell phone is a small device that can be used to make phone calls and send text messages on the go, adding the word "smart" to a phone can be confusing, aren't all phones smart? [4] A smart phone is sometimes called cell phone, because it can make calls but not vice versa [4]. A smartphone can be considered a miniature computer that has a virtual store of many applications such as games, different browsers, maps, emails, image editors, and that help to turn it into a device that is smarter than a regular cell phone [4]. Authors in Ref. [5] define a smartphone as "a next generation, multifunctional cell phone that provides voice communication and text messaging capabilities and facilitates data processing as well as enhanced wireless connectivity" [5]. According to Ref. [5], a smartphone could be considered a combination of "a powerful cell phone" [5] and a "wireless-enabled PDA" [5].

A smartphone has many additional features compared to a regular cell phone such as a color LCD screen, wireless capabilities, that is Wi-Fi, Bluetooth, infrared, etc., a large memory and a specialized operating system (OS) with an offer of many downloadable applications [5]. The emerging new technologies stated above available in smartphones along with the different new applications existing in the market, made of the smartphone a personal device that is not always on, but is always somewhere on us providing a ubiquitous and pervasive computing environment full of seamless services and applications that has most changed our lives [6]. As stated by Romero J. in [6], the smartphone helps users to get the required information whenever needed and to stay connected any time and at any given location [6].

The difference between a smartphone and a cell phone is mainly due to advances in three areas, which are hardware, that is, high-resolution screens, keyboards, cameras, processors, sensors, software, that is, operating systems and various supported applications, and network infrastructure, that is, 3G and 4G networks, and an increasing wireless bandwidth that allows the applications to offload data storage and processing to the cloud [6]. Equipped with different sensors, the smartphone world is considered different: for example, using the smartphone's accelerometers, basic health indicators can be followed, and using the GPS, traffic patterns could be monitored [6]. Many applications for augmented reality were also developed allowing, for example, to point your phone at a restaurant and see customer reviews about it. As stated in Ref. [6], smartphones are considered to become a "sixth sense" for the user, allowing a variety of functionalities.

3. Internet of Things (IoT)

3.1. Definitions

The Internet of Things, also shortly known as IoT, is a term consisting of two words: the first word "Internet," which is "a network of networks and a global system of interconnected computer networks that use TCP/IP as a standard Internet protocol (IP) to connect millions of users and multiple private, public, academic, business, and government networks" [7]. The second word "Things" consists of any real-world object such as home appliances, clothes, etc. or living things such as plants, animals, and people [7]. The term "Internet of Things" was invented by Kevin Ashton, Executive Director of the Auto-ID Center in MIT, in 1999 and its definition varied among academicians and researchers [7]. The best definition of IoT would be according to Ref. [7]: "An open and comprehensive network of intelligent objects that have the capacity to auto-organize, share information, data and resources, reacting and acting in face of situations and changes in the environment" [7]. IoT aims at providing the vision of "enabling anytime, anywhere connectivity for anything and not only anyone by providing unique identity to each and every object" [7]. In the deployed IoT networks, sensors are attached to physical objects and keep track of their data, to allow their tracking on the Internet [7].

There exist many aliases for the IoT concept; these include *"Internet of Objects," "web of things," "connected devices,"* and *"technology omnipotent," "omniscient," "omnipresent," "web of things,"* and *"embedded intelligence."* IoT should not be confused with other terms such as ubiquitous computing where "technology becomes virtually invisible in our lives" [7], pervasive computing in which "virtually every object has processing power with wireless or wired connections to a global network" [7], cyber physical systems, which "helps bringing the real and virtual worlds together" [7], machine-to-machine interaction in which "devices are communicating end to end" [7], human–computer interaction, which "concerns the design of interaction between people and computers" [7], and ambient intelligence, which is "a developing technology that will make our lives responsive and environment sensitive" [7].

3.2. IoT structure

The IoT is a global network connecting things through numerous technologies such as RFID and barcodes to name a few [8]. The International Telecommunications Union (ITU) has structured the IoT into the following four dimensions: (1) tagging things, (2) feeling things, (3) shrinking things, and (4) thinking things [8]. In *tagging things*, RFID tags are used to automatically identify and track the attached object. In *feeling things*, sensors are used to collect data from the physical environment such as temperature, pressure, etc. [8]. In *shrinking things*, nanotechnology is used for tiny things: for example, "the use of nanosensors to monitor water quality" [8]. In *thinking things*, the smart things need, in addition to communication, to process information, make self-maintenance, and make independent decisions; this vision changes the way of information communication from human-human to thing-thing [8]. The structure of IoT is better illustrated in the following **Figure 1** [9].

3.3. IoT technologies

3.3.1. Radio-frequency identification (RFID)

RFID is a wireless identification technology that uses radio waves to identify an object or a person [7]. The first use of RFID was during the second world war to identify friend or foe aircrafts in 1948. The technology was later on founded at the Auto-ID center in MIT in 1999 [7]. The RFID systems consist basically of three elements: the *RFID Tag* serves to uniquely identify the attached object and carries data about it, the *RFID Reader* is the equipment used to power

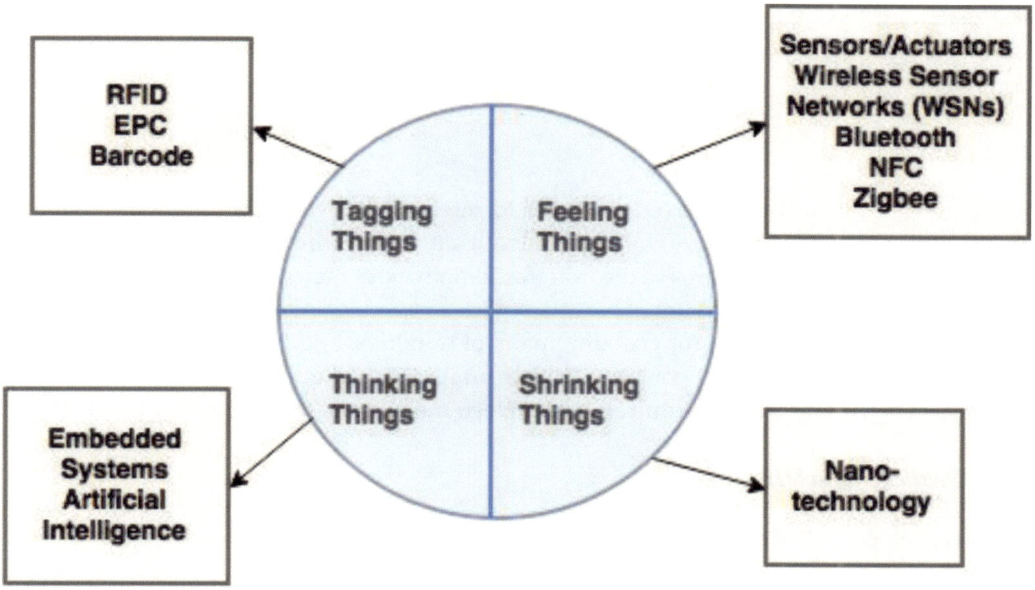

Figure 1. IoT structure in four dimensions and example technologies.

the tag, and read/write data to the tag [7]. The data read by the RFID reader from the RFID tags in its vicinity need to be processed then by a software system, known as the RFID middleware, which is the third component of an RFID system [10]. The *RFID middleware* serves to manage readers, filter, format, and process raw RFID data captured by the tags and send the processed data to the various interested backend applications [10, 11]. There exist three different versions of RFID tags depending on the power supply: *passive tags, active tags*, and *semipassive* [10]. Tags can be also classified based on their type of memory, for example, *read-only, read-write*, or *write-once and read-many*. RFID tags use the ISM (industrial, scientific, or medical) frequency ranges and have three types of frequencies: low frequency (LF), high frequency (HF), ultra high frequency (UHF), and microwave frequency [10]. RFID technology is cost effective, is considered very important in IoT networks for helping with tracking and identification of objects, and is used in a very broad range of application areas [10, 7].

3.3.2. Electronic product code (EPC)

Developed by the AutoID center in 1999 in MIT, the EPC code (64/98 bits) can store information about the unique serial number of a product, its specifications, and manufacturer's details [7]. The EPC has four components which are "object naming service (ONS)," "EPC discovery service (EPCDS)," "EPC information services (EPCIS)," and "EPC security services (EPCSS)" [7]. The EPCglobal Network [7] was created by the EPCglobal Organization to share EPC data and is a framework consisting of the ID System; EPC tags and readers; EPC middleware, which takes care of basic data formatting; EPC Information Services, which enable exchange of information between partners; and Discovery Services, which enable users to get and search for EPC data [12]. The EPC was basically designed to be stored on an RFID tag to identify a specific item and its associated data such as origination point and date of production [12].

3.3.3. Barcode

The barcode system uses a barcode scanner to interpret the value in the barcode label to obtain a unique code that is used for object identification [7]. Barcodes are "optical machine-readable labels attached to items to record information about them, and they are usually read by laser scanners" [7]. Three types of barcodes exist: *alpha numeric* used for encoding numbers and characters, *numeric* used for encoding pairs of numbers, and *2-dimensional*, which looks like squares or rectangles that contain many small, individual dots [7]. The disadvantage of this system is the line of sight requirement between the barcode and the reader [10, 13].

3.3.4. Internet protocol (IP)

Internet protocol (IP), developed in 1970s, is considered main network protocol used for sending packets in the Internet [7]. There are two different versions of IP addresses: IPv4 (default version) and IPv6. IPv6 was developed to extend the number of available and supported IP addresses [7].

3.3.5. Wireless fidelity (Wi-Fi)

Wi-Fi (wireless fidelity) allows devices to communicate over a wireless signal, and contains any type of wireless local area network (WLAN) device supporting any of the following IEEE 802.11 specification versions: 802.11a, 802.11b, 80.2.11g, and 802.11n [7, 14]. Today, Wi-Fi is delivering the high-speed WLAN in-building connectivity to hotels, homes, airports, and cafes through the use of wireless access points (APs) [14]. Although encryption is considered optional in Wi-Fi, three techniques of encryption have been defined and applied to Wi-Fi to ensure security. These techniques are "wired equivalent privacy (WEP)," "WiFi protected access (WPA)," and "*IEEE 802.11i/WPA2*" [14]. To access a Wi-Fi network, Wi-Fi enabled devices (e.g., laptops) are needed, which can communicate wirelessly in any Wi-Fi equipped location [14].

3.3.6. Bluetooth

Bluetooth is a cheap communication technology, deployed for small distances (10–100 meters), that allows connection between devices, for example, laptops, PDAs, smartphones, printers, cameras, etc. without the need for cabling. Bluetooth is considered the main technology for creating a personal area network (PAN) to share data such as text, images, videos, and sounds, and it uses the IEEE 802.15.1 standard. Bluetooth allows users instantaneous connections between several devices and assures protection against interferences and safety in the sent information [15]. The Bluetooth technology operates in the ISM band, which is 2.45 GHz. Some standard Bluetooth applications include but are not limited to communication between hands-free device and a mobile phone or car radio, and transfer of files between devices [15]. Bluetooth can have some security risks, because it is an open system. Security can be implemented at the level of devices and services [15].

3.3.7. ZigBee

ZigBee technology was created and developed in 2001 by the ZigBee Alliance [7]. As defined in [16], ZigBee is "a low power, low cost, low data rate, and short range (around 100 meters), wireless network protocol based on the IEEE 802.15.4 standard" [16]. ZigBee is widely used in home automation, medical monitoring, industrial controls, and digital architecture [7]. ZigBee was developed to enhance the features of wireless sensor networks (WSNs) and is widely used for wireless home automation systems (WHASs); however, it has many related challenges such as resource constraints, low memory, limited battery, limited processing power, limited range, technological limitations related to the IEEE 802.15.4 standard, interferences with systems operating in the same free band, and internet connectivity, which is needed in WHAS for remote monitoring [16].

3.3.8. Near-field communication (NFC)

NFC is a short-range (theoretically 20 cm, but in most scenarios, typically 4 cm) wireless technology developed by "Philips and Sony companies" that works at the 13.56 MHz frequency and allows customers around the world to easily make transactions, connect electronic devices, and exchange digital content [7]. NFC technology is compatible with existing RFID infrastructure and contactless smart cards and uses the same standards such as ISO/IEC 14443 standard, which is one of its biggest advantages [17]. NFC has an easy and simple connection method, can work also in dirty environment, and does not require a line of sight for reading and executing transactions [7]. Some example applications of NFC include but are not limited to mobile payment such as Google Wallet, and mobile ticketing such as Oyster Card [17].

3.3.9. Wireless sensor networks (WSNs)

A WSN consists of hundreds to thousands of "sensor nodes" communicating with each other and passing data related to either "physical" or "environmental conditions" such as pressure, temperature, motion, location, sound, etc. [7, 18]. WSNs are used widely in IoT for many types of applications such as agriculture monitoring, patients monitoring, military applications, highway monitoring, civil and environmental engineering applications, forest fire, industrial automation, home control, building automation, etc. [18]. An example of the use of WSNs in healthcare is the use of sensors to monitor medication responses from a patient's body [7]. A key issue faced when designing a WSN network is energy efficiency, that is, designing for a "long network lifetime and limited network maintenance and deployment costs" [10]. A middleware system is needed to provide the multiple services required by WSN applications and allow for scalability, power saving, and quality of service (QoS) while designing WSN applications as suggested in [10].

3.3.10. Actuators

As defined in Ref. [7], an actuator is "a device that actuates or moves something; converting energy into motion or mechanical energy" [19, 7]. Typical applications of actuators are implemented in the industrial and manufacturing fields [7]. There are three types of actuators: *electrical actuators* are "AC and DC motors, stepper motors, and solenoids" [7]; *hydraulic actuators* "use hydraulic fluid to actuate motion" [7]; and *pneumatic actuators* "use compressed air to actuate motion" [7].

3.3.11. Artificial intelligence (AI)

Intelligence has been embedded and hidden in the network connected devices that help people to ease their daily activities [7]. AI refers then to "electronic environments and devices that are sensitive and responsive to people's presence and activities" [7]. AI is considered "*embedded,*" because the devices used are seamlessly embedded within people's environment, "*context-aware*" because these devices are used to know people's situations and context conditions, "*personalized*" because they can be customized to the needs of users, "*adaptive*" because it changes depending on the users' needs, and "*anticipatory*" because it can predict the user needs without conscious mediation [7].

3.4. IoT architecture

A reference model has not yet been suggested for the IoT architecture, although there are an ever-increasing number of proposed architectures for this new trend such as the ones described in [20–24]. Among the most common architectures of IoT we find the 5 layers model described in [25]. The first layer of this model is named the *objects layer/perception layer* and represents the physical objects, for example, sensors, actuators, etc. of the IoT that serve to collect information using standardized plug-and-play mechanisms to serve the heterogeneous devices [25]. The second layer is the *object abstraction layer,* which transfers the collected data from the objects layer to the service management layer using various technologies such as RFID, 3G, 4G, Wi-Fi, Bluetooth, and handles data management processes and cloud computing [25]. The third layer is the *service management layer,* which is the middleware layer that processes the received data, delivers the processed data and services to the interested applications over the network, and makes decisions [25]. The fourth layer is the *application layer* and is the one responsible for providing the requested smart services to the customers or connected applications that meet their needs in the various domains such as healthcare, transportation, and industrial automation [25]. The fifth layer is the *business layer,* which supports decision-making processes based on big data processing and analysis, manages all the underlying four layers of the IoT architecture, and enhances the services provided to the users and maintains their privacy [25].

In Ref. [10] we proposed and developed a flexible middleware solution architecture that has five layers and is compatible with the IoT architecture discussed above. We developed the proposed architecture more and applied it to e-health in [26]. The FlexRFID middleware in [10] serves getting data from the heterogeneous automatic identification devices and sensors, processing them, applying the business rules specified by the connected applications, and disseminating the processed data to the interested applications. Our middleware, that is, FlexRFID was tested with multiple application domains, such as smart library management [27], supply chain management [28], and healthcare scenarios [29].

4. Internet of Everything (IoE)

The Internet of Everything (IoE) concept is a fairly new concept that was developed by Gartner in 2015, and there is still confusion about the difference between IoE and IoT [30]. The IoE as defined in Ref. [30] is "bringing together people, process, data, and things to make networked connections more relevant and valuable than ever before-turning information into actions that create new capabilities, richer experiences, and unprecedented economic opportunity for businesses, individuals, and countries" [30]. IoE definition means "connecting people in more relevant ways, converting data into intelligence to make better decisions, processing this data and delivering the right information to the right person at the right time, and connecting things which denote any physical devices or objects connected to the Internet or to each other for intelligent decision making" [30]. In other words, IoE describes an environment where many objects are identified, sensed through the use of sensors to detect their status and measure their conditions, and connected over public/private networks using specific standard/proprietary protocols [30]. The IoE is a term describing the intelligence added to every device in order to give it some added functionalities, the device could be any of the following: smartwatches, smart appliances, smart beds, health monitoring devices, smart connected cars, and others [31]. The difference between IoE and IoT is that IoE consists of four parts: "people," "process," "data," and "things" and builds on top of IoT, which consists of one part, which is "things" [30].

As reported by Cisco "IoE is capable of helping organizations achieve many public-policy goals, including increased economic growth and improvements in environmental sustainability, public safety and security, delivery of government services, and productivity" [32]. Also, Cisco reported in [32] the five drivers of IoE value for the public sector which are (1) *"employee productivity"* consisting of "improved labor effectiveness for new and existing services" [32], (2) *"connected militarized defense"* consisting of "improved situational awareness and connected command centers, vehicles, and supplies" [32], (3) *"cost reduction"* consisting of " improved labor efficiency and reduced operational costs" [32], (4) *"citizen experience"* consisting of "shorter search times; improved environment; better health outcomes" [32], and (5) *"increased revenue"* consisting of "improved ability to match supply with demand; improved monitoring and compliance" [32]. Cisco CEO, John Chambers, believes that "IoE will have a dramatic impact on everything from city planning, first responders, military, health, and dozens of other environments" [32].

IoE is believed to "extend the IoT emphasis on machine-to-machine (M2M) communications to describe a more complex system that also encompasses machine-to-people (M2P) and technology-assisted people-to-people (P2P) interactions" [2]. IoT and M2M are often considered synonymous and sometimes used interchangeably; however, IoT refers to "connection of systems and devices to the broader Internet" [2, 33].

5. Application areas of IoT and IoE

5.1. Applications based on the IoT

Information generated and communicated by the enabling objects in IoT can drive many possible applications in many domains such as supply chain management (SCM), transportation, healthcare, and environment and disaster monitoring, etc. [9].

5.1.1. Logistics and supply chain management (SCM)

In IoT society, many logistics applications have been developed to track movements of goods in real time using the different technologies discussed above, such as the systems reported in [34–36]. The data scanned from the RFID tags, barcodes, NFC, and mobile phones were transmitted to the logistics center, and then transmitted through diversified transmission protocols such as WSNs, GSM network, 3G, 4G, or even 5G network to be processed [9]. Some example applications of the IoT in logistics and SCM as reported in [9] include *Supermarket chain management* [34], which tracks goods in real time using WSNs, barcodes, and RFIDs, and controls automatically the stock; *Aspire RFID* [37], which is "a middleware with a range of tools to facilitate RFID deployment, in addition it uses the session initiation protocol (SIP) to detect the location and mobility management of RFID tags" [9], *logistic geographical information detection UIS* [38], and others. The use of sensors in SCM provides rich data about supply chains and also on conditions and location of goods in real time [39]. This helps supporting "circular economy," because tracking a product from manufacture to recycling helps enabling new ways for resource optimization [39]. To guarantee an efficient implementation of IoT, applications in SCM should ensure some basic capabilities such as *autonomous control* by having small decentralized control units [9], *smart logistics entities* by using sensors to track items and protect them from thieves by triggering alarms when a set of conditions is met [9], *unique addressability* by using a set of technologies such as RFID and WSNs to help tracking "the right product, right quantity, at the right time, in the right place, satisfying the right conditions, and having the right price" [9], and *enterprise resource planning* (ERP) interface to help communicate to the customer the right information about the products [9].

5.1.2. Transportation

IoT is considered to have many advantages for solving the numerous challenging transportation problems, and many applications such as the *road condition monitoring and alert system* reported in [40] were developed to communicate the road conditions in real time and alert their users about any congestions or existing problems such as accidents [40]. Other applications such as *license plate identification* as reported in [41] have been implemented to solve the problem of finding parking spaces and securing the vehicles [41]. *Electric vehicles* have also been supported by governments in many countries "to reduce the fuel cost and the impact of global warming"; systems such as the one in [42], that is, remote performance monitoring system and simulation testing, have been designed "to monitor the performance of lithium-ion (Li-on) batteries for electric vehicles" by using WSNs to report the route's status to the drivers and help them save their vehicles' batteries. Using IoT nowadays, many electric vehicles' manufacturers offer applications that can remotely monitor the vehicles' batteries power and schedule their charging [39]. As reported in [39], in the future, fully autonomous vehicles are expected to be integrated in a smart transportation system, and a trial system has been implemented in Newcastle that gives signals to drivers about when to adjust their speed if traffic lights are about to change [39]. Also parking sensors have being tested in Milton Keynes [39]. IoT has been also used at London City Airport to improve customer experience and passenger flow through the use of sensors deployed throughout the airport that send data to passengers' smartphone applications to help them order from shops and know about queue times [39]. Other systems such as transport vehicle monitoring system based on IoT in [43] uses GPS, RFID, and 3G/4G technologies to monitor and administer the status of goods in real time [43]. IoT is also used nowadays in vehicular ad hoc networks (VANETs) and is driving the evolution of Internet of Vehicles (IoV) paradigm. In conclusion, the IoT-based applications in the transportation field should at least include the following units as suggested by authors in [9]: a *vehicle system* equipped with GPS and wireless communication technologies [9], the *station system*, which is "responsible for receiving data from the monitoring center and displaying real-time transit vehicle information" [9], and the *monitor center*, which is "responsible for comparing the received real-time data with events in the database and integrate the road traffic information for visualization" [9].

5.1.3. Healthcare

Many IoT solutions were implemented to improve human health and well-being and facilitate access to healthcare in rural areas such as the one described in Ref. [44]. The solution in [44] is based on RFID data communicated by active RFID tags worn by people who register with the rural healthcare center (RHC). The RFID tags are used to continuously monitor and control the patients' healthcare parameters such as temperature, blood pressure, etc., detect any change in them, and communicate them to the RHC doctor. IoT-driven healthcare systems and technologies can be used for prevention and early identification of diseases [39] and are basically used for hospitalized patients whose status requires continuous monitoring and attention, or for monitoring an aging family member at home [47]. Other examples of applications include the integration of a variety of devices in the patient's environment such as the use of smartphones to monitor vital signs and transmit health data directly to the care centers [45]. Some advanced systems such as the one in [46], that is, noncontact health monitoring system (NCHMS) uses classification and recognition-based algorithms and equipment equipped with cameras and microphones to analyze the user's facial expressions and detect any anomalies. In conclusion, the IoT-based applications in the healthcare field should at least include the following units as suggested by authors in [9]: *Tracking and monitoring* using any wearable WSN or RFID devices to generate and communicate health vital signs, *remote service* such as telemedicine and home diagnosis [47], which is necessary to provide emergency help to patients suffering from critical illnesses, *information management* used to manage the large amounts of data produced and captured about a patient such as medical history of medications and allergies, and *cross-organization integration*, which ensures an integration and communication among the hospital information systems, the patients' homes, and other medical care centers [9].

5.1.4. Environment and disaster

IoT technologies are used nowadays to minimize the effects of natural disasters by providing alerts and helping in the disaster recovery process [9]. Many examples of systems exist in the literature for environment monitoring such as "Health Monitoring and Risk Evaluation of Earthen Sites (HMRE2S) model" suggested by Xiao et al [48]; which collects "temperature, humidity, and light information to evaluate the healthy level of the earthen sites by applying the concept of artificial antibodies to identify unusual environmental factors" [48], and "*Smart heat and electricity management transportation*" suggested by Kyriazis et al. [49] that "uses smart meters for electricity consumption and mobile sensors to assess the effect of real-time electricity usage on the energy consumption of buildings and individual appliances, etc." [9]. In conclusion, the IoT-based applications in the environment and disaster field should at least include the following components as suggested by authors in [9]: *environment sensors*, which help gathering and processing information such as humidity, temperature, and pressure from the environment, *WSN and mobile communication* (3G and 4G) helping to communicate the sensed information to other users or systems and trigger the necessary alerts, and *participatory sensing applications*, which, by the use of multiple sensors and devices to capture the environment data and sense the physical world, and to help making the right decisions when facing a disaster, for example [9].

5.1.5. Smart home and smart buildings

Home automation can be made possible using IoT technologies to allow us to remotely control our home's appliances based on our needs [50]. Example applications include but are not limited to monitoring of utility meters, energy, and water supply to avoid overloading or leaks, and gardening sensors, which could be used to water plants according to their needs and measure their vitals such as light, humidity, and moisture [50]. Connected to the IoT, smart buildings' energy and maintenance could be optimized and predicted, along with increased comfort, security, and safety for the building users [39].

5.1.6. Smart agriculture

The IoT technologies, such as field-based sensors, can be used to monitor soil humidity, moisture, and nutrition, automatically adjust the temperature to maximize agricultural production, and communicate with weather stations to get the latest forecasts [50, 39]. They can also help for an accurate fertilization and watering [50]. Sensors used for animal tracking help in monitoring livestock for disease and accidents, and providing better opportunities for husbandry [39]. "Smart farms" may also share data with other farms, consumers, and regulators [39]. The major opportunities provided by IoT for agriculture are maximizing yields by automatically identifying damaging weeds and reporting their location to farm owners or autonomous weeding machines, improving food traceability by tracking food and informing consumers about their provenance, origin, and production methods, and tackling environmental challenges such as the use of 3D accelerometers to detect injuries in cows and monitor them within the livestock, which allows for an early adoption of preventive measures [39].

5.2. Applications based on the IoE

The IoE has been used to help "automated and people-based processes" by extracting and analyzing real-time data from the millions of connected sensors [51]. IoE has been also used for environment sustainability, public policy goals, and economic goals [51]. The use of IoE has been facilitated by the expansion of cloud computing, which helps connecting everything online [51]. "Smart cities" will benefit from IoE to address city-specific concerns along with big data processing, for example, using sensors in monitoring highways and traffic, education, healthcare, agriculture, and environment [51]. These cities will most likely enhance the living conditions of citizens in the future by forming "Smart + connected communities" [51]. IoE will be considered a critical element in implementing new features of the future cities such as *smart grid* and traffic control [51, 52]. According to Cisco, "cities stand to benefit the most from IoE related projects, implementations and platforms," which helps providing real-time, context-specific intelligence and analytics to serve the city's specific needs [53]. Many examples of how Cisco was involved in developing new models for cities have been included in [53]; however, there exist many challenges for the IoE-enabled cities such as the need for new operating models, coherent IoE deployment plans, new ways to preserve the cities assets such as data, new governance models, and the need to face societal challenges such as pollution, and CO2 emissions [53]. IoE technology architecture for cities is suggested in [53], which is a multilayered architecture that provides handling millions of devices and sensors, processing and streaming of big data and decisions, storage and analytics of data, and APIs for adding new services or applications [53].

IoE is also expected to ensure safety in the mining industry of fossil fuels [53]. Another use of IoE is in the educational sector where it facilitates access of students to E-learning and M-Learning, and provides more feedback and progress monitoring [51]. As reported by Zielinski [52], "The IoE provides a new business model for companies, which ultimately implies lowering the cost of energy distribution, automate billing and service calls as well as providing proactive response to environmental condition" [52].

In an IoE world, we can find multiple applications integrated in multiple ways: for example, public and private organizations usually integrate IoE applications with their existing solutions such as ERP, SCM, CRM, human resources, etc. [54]. This high level integration allows for better service guarantee and higher security [54]. IoE solutions are also expected to access data from a single-purpose device initially, an example of this is connected automobiles running multiple applications such as location detection, emergency calls, etc. [54].

6. The use of smartphones in IoT and IoE

The technology necessary for all the example applications of IoT and IoE, stated in the previous sections, to succeed is available today. RFID, Bluetooth, NFC, 3G, 4G, 5G, etc. can transfer data over the Internet, also batteries' technologies have evolved; for example, wireless and solar power batteries and long-lasting batteries are available in today's market [3]. In addition, different types and sizes of sensors exist and can be used to monitor various industrial processes [3]. The challenge is to include all the aforementioned technologies into one light, inexpensive, user-friendly, multipurpose, and portable device that can be easily used by people in their daily lives [3]. Such a device is today an existing reality which is the smartphone. The Smartphone is equipped with a range of built-in sensors such as accelerometers, motion sensors, position sensors, and environmental sensors, that is, barometers, thermometers, and photometers measuring pressure, humidity, ambient temperature and illumination levels, etc. Some other kinds of special sensors measuring health vital signs, such as body temperature, ECG value, blood glucose level, stress level, body fat percentage, heart rate, etc., can be integrated into the smartphone [3]. All these sensors produce large volumes of data in structured form such as GPS or acceleration data and unstructured form such as pictures or videos [55]. The smartphone's cameras and microphones are also used to detect and record images in many smart applications used in IoT [3]. The smartphone is also equipped with a variety of connectivity technologies such as NFC, Bluetooth, Wi-Fi, and cellular, which allow it to connect and interact with other devices and sensors and be the brain of the IoT world [56]. An example of the use of IoT-enabled smartphones given in [3] is "traffic congestion control on specific roads using Google Maps; data are automatically being collected from users' smartphones moving along a specific road at a specific time, processed and sent to all connected users to Google Maps interested in getting this information" [3]. Another example is the use of smartphone to open a smart door of a hotel room in some parts of the world, once you approach it. This could be extended to office access control or garage access door opening [56].

Smartphones along with other IoT devices will play a main role in the expansion and use of this new terrain of Internet of Things [55]. The smartphone is considered to be "at the heart of a growing universe of connected devices and sensors" [55], also the rise of the smart wearables such as Apple Watch and Android Wear plays an additional role in creating an intelligent body area network (BAN) for the user where he stays connected most of the time [55]. The NFC technology integrated to the smartphones allows them also to act not only as sensors but as actuators triggering many actions such as payments, TV control, cars control, and home automation [55].

As reported in Ref. [55], smartphones can be used in an IoT setup along with four application categories: (1) *Personal IoT* where we find an increasing number of applications targeting health and fitness, and helping to solve everyday problems for users; (2) *group IoT* where smartphones can be used in the context of connected cars to check the system status, or in smart homes; (3) *community IoT* where crowd-sourcing applications could be used by citizens to contribute to a smart city; and (4) *industrial IoT* where smartphones are used for business to consumer (B2C) purposes such as sending customized services and vouchers in real time [55]. Future applications of the use of IoT through the smartphone include viewing data and controlling sensors anywhere; for example, at home or in a workplace, the smartphone could be used to "control a smart air conditioning or an alarm system at home from an application, or technicians may be alerted on their smartphones when a factory machine at a customer site is overheating and probably needs attention" [56]. In a smart city context, a smartphone application could be used to check the queue in a local store and see whether an item is in stock in real time, reserve the item, and call the client service for delivery. Other applications could be used in the smartphone to get data about noise and traffic congestion in the city to improve the residents' living experience in the controlled area.

Authors in [58] propose a four layers model named "*k-Healthcare*," which is considered a comprehensive platform for accessing patients' health data using the smartphones' sensors and applications. The model in [58] is composed of the following layers: (1) the *sensor layer*; which consists of sensors used to detect the patients' vital signs such as blood oxygen and pulse and the smartphone built-in sensors such as barometers, temperature, and humidity sensors, along with RFID tags used to for objects identification. All of these IoT devices are used by the k-Health platform to get data and send them to the other layers for further processing [58]. (2) The *network layer* is "the communication layer that connects the IoT devices with WAN using different protocols such as 802.16 for 3G, IEEE 802.16m for 4G, IEEE 802.20, ZigBee, etc." [58] (3) The *internet layer* is responsible for data management and storage using cloud storage or physical storage. Finally (4) the *services layer* "provides direct access of data to patients and professionals such as doctors, hospitals, and medicine supply chains using various protocols such as HTTP, HTTPS, web services, etc." [58]

Using our FlexRFID middleware discussed in the IoT architecture section, **Figure 2** shows the use of smartphones at the different layers of the IoT architecture.

From **Figure 2**, it is clear that the smartphone could be used as an automatic identification and sensing device at the level of the *sensing/auto-tracking layer* and as a backend device at the level of the *application layer* where different users accessing different applications could get the needed services.

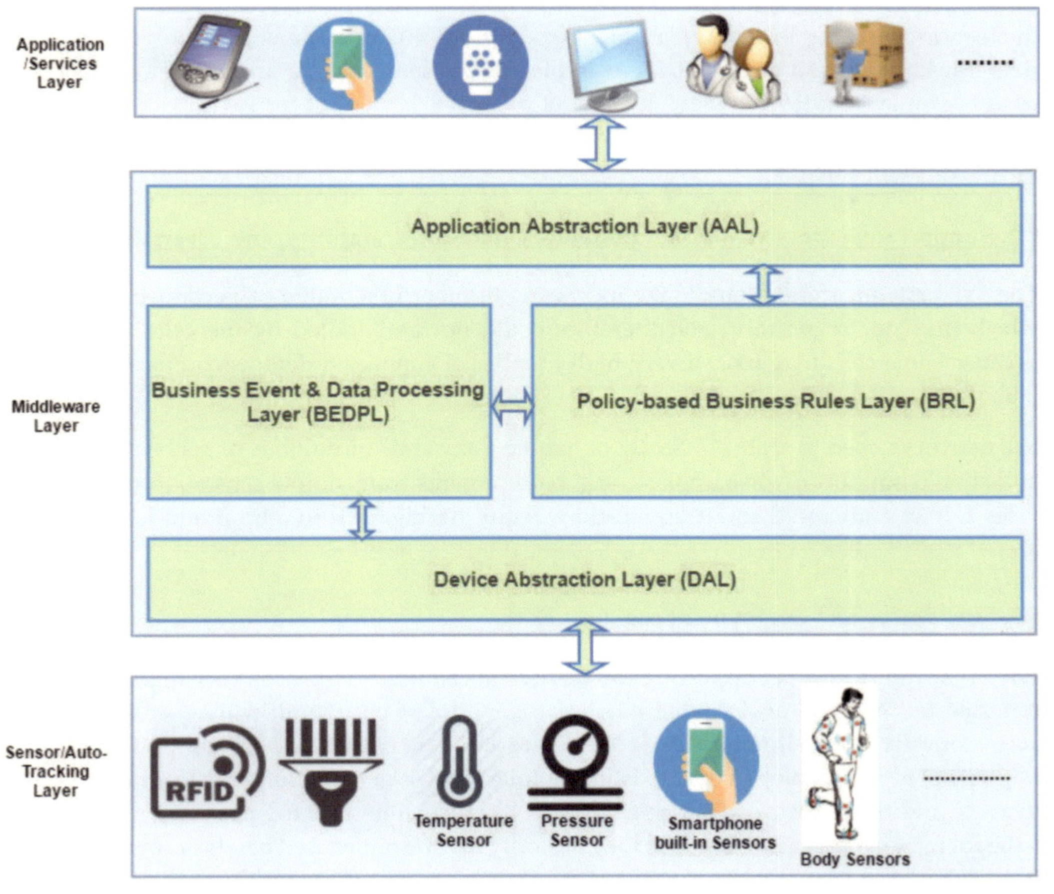

Figure 2. The use of smartphone in FlexRFID middleware.

7. IoT networks versus mobile cellular networks

Regular cell phone networks fall short for IoT requirements, basically for battery life, cost, and wireless coverage. That is why many wireless carriers around the world are building new cellular networks to work with current and upcoming IoT devices and solving one or all of the three new IoT requirements [59]. For example, Orange and SoftBank are building nationwide IoT networks, Vodafone is upgrading its networks, and Cisco and Samsung are inventing and selling new devices to expand the IoT concept [59]. A comparison between cellular networks and IoT networks concerning these perspectives is given below.

7.1. Long battery life

Mobile cellular networks were designed to coordinate moving from one cell to another, called "hand-off" mechanism, without interrupting a phone call by using sophisticated algorithms [59]. To ensure this, mobile cell phones should communicate multiple times per second with the cell tower, which is very expensive in terms of battery consumption. In order to save the battery power for years, the IoT new cell networks' devices should spend most of their time in sleep mode using low power radio chips and optimized to minimize the power cost of data transmission and reception, for example, to read sensors' data or activate a control such as an alarm system [59]. Achieving years of battery is important for IoT devices, because it eliminates installation costs and that is why new networks need to be built to save the battery power consumption [59].

7.2. Support for a massive number of devices, network scalability, and diversity

The IoT network should handle the increasing number of simultaneous connected devices, which may not be uniform and therefore could not be handled by the cellular network, because some cells may have a very high number of connected devices compared to others [60].

IoT networks need to scale efficiently to handle thousands or millions of connected devices, and should support diverse applications' requirements from simple sensors to tracking services to more advanced smart applications requiring higher throughput and lower latency [61].

7.3. Low device and deployment cost

The opportunity cost of supporting IoT devices in cellular networks is very high, because as reported in [59], "IoT devices that pay less than a dollar per month will never get network access priority over cell phones with $100 voice and data plans" [59]. IoT connectivity can be implemented and deployed over existing cellular networks using software upgrade in order to avoid additional costs of acquiring any new hardware [60]. The IoT networks are designed to be robust to interference, because they basically use the unlicensed bands or unused "guard bands" between the channels of licensed cellular spectrum, which are cheap compared to the licensed bands [59].

http://dx.doi.org/10.5772/intechopen.69734

7.4. Extended coverage

IoT networks should also handle coverage concerns that are not covered by cellular networks in places such as basements of buildings, underground parking lots, and rural fields [60]. This extended coverage is required by many IoT applications to get the necessary data from the deployed sensors and send them in real time to the interested applications [60]. The IoT networks handle this by maximizing deep indoor penetration rather than bandwidth, and through the use of self-deployable gateways that could be installed like Wi-Fi routers [59].

8. Future opportunities and challenges of IoT and IoE applications

8.1. Future opportunities

People are getting more connected and devices are becoming smarter, and new network architectures are adapting to this: such as big data, cloud infrastructure, and mobility, which are important parts of the Internet of Everything movement. In the future, every device is considered to be communicating to some extent and this will have an impact on the growth of cloud services [57]. The IoE will reinvent industries at three levels in the future; the first level is *business processes and services* will be improved by the new trends in digital technology [30], the second level is *business model*, which will be changed as new ways of doing business industries will emerge and companies will tend to digitalize more its products and processes, an example stated in [30] is that of Nike with its connected sporting clothes in the healthcare domain [30], the third level is *business moment*, which is the need to compete with other businesses [30]. Also the IoE will generate large volumes of data in real time, and therefore, businesses will need big data, storage, and analysis tools to manage these data; generate high-level information and services; and turn them into money. As millions of objects, sensors, devices, and people get more connected and collect more data; a critical task for companies will be to tackle the issues of privacy and security that arise through the use of IoE technology [30].

8.2. Challenges

IoT and IoE offer numerous revolutionary benefits to consumers in many areas such as healthcare and supply chain management (SCM), to name a few. The use of connected medical devices, for example, can engage patients in their own care and allow doctors to respond in real time and better manage the patients' diseases. Despite these opportunities in many application fields, security and privacy risks arise due to the increased connectivity among devices, people, and the Internet. According to Ref. [62] IoT technology presents a higher potential of security risks at different levels such as enabling unauthorized access to personal information and identity theft, creating safety risks and allowing attacks to other connected systems; for example "security vulnerabilities in an IoT device could be used to launch a denial of service attack on the consumer's network to which it is connected, this device could also be used to send malicious emails and messages to other devices" [62]. Unauthorized persons might also create physical safety risks by exploiting the security vulnerabilities of IoT devices: for example, a hacker can change the settings of an insulin pump to no longer deliver insulin to the concerned patient, which creates health problems and crisis [62].

Companies experiencing the IoT technology may not have enough experience in dealing with the security issues stated above and therefore find securing IoT devices and communications a challenging task [62]. Also the structure of some IoT devices is sophisticated and the manufacturers find it difficult or expensive to apply a security patch in them if a specific vulnerability is discovered [62]. In addition, some IoT devices are made disposable after purchase and therefore, the consumers are often left with vulnerable devices shortly after their purchase in most cases [62].

In addition to security risks, there are many privacy risks involved with IoT such as the collection of sensitive personal daily information such as health information, geolocation, and account numbers and sending data through the cloud [62]. The collection of this information over time could be misused and can help intruders infer future values. Privacy principles state that users should control their personal data and choose the smart environment and technology that protects their private lives [63]. Users usually have difficulty knowing about the existence of IoT devices in their environment, what information is being disclosed and sent in the network, and which parties benefit from this information. Also manufacturers are interested in building services around the collected data rather than selling the devices themselves [63]. According to Ref. [62], researchers state that the smartphones could be used to disclose the user's personality type, demographics, stress level and mood, happiness, etc. [62]. Another privacy risk is that an intruder could intercept unencrypted IoT data remotely while sent in the IoT network, combine, analyze, and act upon them [63]. The above security and privacy challenges may result in an undermined consumer confidence and a decrease in the IoT technology widespread adoption, which will surely affect the overall societal acceptance of IoT services [62].

Our proposed middleware architecture called FlexRFID tackles the security and privacy issues in the IoT environment at the application level by using policies as described in [29]. These policies allow the applications to specify the security, access control and privacy rules that should be applied on data before getting them, and therefore minimize the possibilities of compromising user's sensitive data. At devices level, new security models other than strong encryption are required in IoT because of the devices' limited capabilities such as limited size, computing, and processing power [63].

Authors in [1] define features of IoT security and privacy in the healthcare field, including security requirements of medical data, which are "confidentiality, integrity, authentication, availability, data freshness, non-repudiation, authorization, resiliency, fault-tolerance, and self-healing" [1]. In addition, the authors in [1] identify challenges for providing secure IoT services, which include (1) the computational, memory, and energy limitations of IoT healthcare devices, (2) multiplicity of IoT devices in healthcare, (3) mobility of IoT devices through different networks having different security configurations, which requires a challenging task of developing a mobility-compliant security algorithm, (4) scalability of IoT devices and their connection to the global information network, (5) IoT devices are connected to multiprotocol networks using a wide range of communication media and a dynamic network topology,

(6) designing a mechanism for dynamic security updates for the various IoT devices, and (7) designing tamper-resistant packages for IoT healthcare devices to avoid extracting cryptographic secrets, modifying programs, or replacing these devices. In addition to the challenges stated above, there are many other issues that need to be addressed concerning IoT services and devices. According to authors in [1], the most important issues are: the need for a unified standardization effort, the need for special IoT platforms and frameworks, targeting cost analysis of IoT-based services, the need to develop new IoT applications as the technology evolves and new devices emerge, the need for a "business model," and the need for "quality of service" guarantees for most IoT services [1].

9. Security and privacy challenges concerning the use of smartphones in IoT and IoE networks

Security and privacy of smartphones in IoT and IoE should be guaranteed to the maximum, because the smartphone is considered the major personal device used in IoT. Threats and attacks on the smartphone and IoT devices can be divided into the following categories as reported in [64]:

- **Resources**: such as GPS, camera, NFC, and other sensors.
- **Data**: such as messages, calls, contacts' list that could be compromised by malicious apps available for free in the app stores.
- **System information** about the smartphone such as identity, location, and Wi-Fi MAC address that could be disclosed without the user permission.
- **Worms**, which are programs that copy themselves to the various devices of a network and can compromise the security of the smartphone.
- **Spyware and malware applications**, which can monitor the users' data without his/her knowledge and send the data to the attacker.

Other smartphones attacks discussed in [64] include "financial malware attacks, network spoofing attacks, phishing attacks, surveillance attacks and network congestion attacks" [64].

Authors in [64] divide security violation into five categories, which are the following: (1) *"breach of confidentiality"* when "an unauthorized person reads and gets access to the data" [64], (2) *"breach of integrity"* when "the attacker reads and modifies the data" [64], (3) *"breach of availability"* when "the attacker destroys and deletes the data" [64], (4) *"denial of service"* when "the attacker attacks the limited resources of the smartphone like filling its memory, draining its battery, etc. and therefore makes it unable to communicate with other IoT devices" [64], and (5) *"theft of services"* when "the resources are used by an unauthorized person" [64]. The five categories of attacks stated above have different effects on the smartphones as major IoT devices, for example, a Denial of Service attack of a smartphone will affect the IoT and the cellular network, a data leakage attack of a smartphone will disclose private data such as online transactions, and a spamming attack will send messages to other smartphones and IoT devices [64].

The study in [64] compares IoT devices and smartphones in terms of many features such as "computation capacity, storage, external storage, authentication, end-to-end communication, expansibility, battery exhaustion, etc." [64]. The study shows that the smartphone has a lot of functionalities and has built-in sensors that allow it to perform most of IoT devices functions [64]. The study also shows the behavior of smartphones in the IoT environment concerning data sharing with other IoT machines, communication with IoT devices and the cloud, supporting more computation in IoT than in the web, and the possibility of sending malicious data to other IoT machines [64].

A survey of more than 5000 consumers from the USA, UK, Canada, Austria, and Japan conducted by Norton in 2016 revealed that some people understand that smartphones and IoT devices present risks and the rest do not care about their information being hacked [65]. As stated in [65] few research studies have focused on the risk of controlling IoT devices by the use of mobile apps installed in a user's smartphone. An intruder can control or get access to the smartphone and therefore control the IoT devices from mobile applications such as control of home appliances and healthcare-sensitive sensors [65]. Mobile applications can send unencrypted sensitive information from a user's phone such as location, call logs, browser history, and account details. Examples of vulnerabilities could be adding browser favorites, downloading and changing call logs, etc. Authors in [65] state the most important best practices that a user can adopt while using IoT devices, smartphones, and mobile apps, which are the following: (1) *using a reputable mobile security app* that identifies potential vulnerabilities before downloading an app, (2) *downloading apps from an official app store,* (3) *being mindful of the app settings* such as apps asking the user to disable security setting that protects installing apps from an unknown source, (4) *keeping the IoT devices current* by installing the latest updates, (5) *protecting the device by choosing a strong and unique password,* and (6) *being stingy with the device* such as protecting the communication between the device and network using an encrypted Wi-Fi connection or a hard-coded LAN connection if available [65].

10. Conclusion

The Internet of Things (IoT) and Internet of Everything (IoE) are rapidly finding their paths in our modern lives, allowing connecting and automating everything around us. This chapter gave an overview about these new trends, their enabling technologies, architecture, and application fields such as smart homes and healthcare. In this chapter, we also talked about the different IoT and IoE enabling technologies available in the smartphone and examples of its use in IoT and IoE scenarios. We proposed a model for IoT implementation that uses the smartphone sensors to sense and transmit data to multiple backend applications using a middleware layer. The applications could be running on a smartphone, which receive the data and present it to the end user, that is, the patient, hospital administration, or physician in the case of healthcare. These data could be stored in special databases or in the cloud and retrieved by the user later on upon need, using the smartphone dedicated application. We also covered the differences between IoT networks and mobile cellular networks in terms of requirements such as the need in IoT networks for long battery life, support for a massive number of devices, network scalability, low device and deployment costs, and extended coverage. Finally, future opportunities and challenges of IoT and IoE have been addressed especially the security and privacy risks of using the smartphone in these networks and possible countermeasures. By addressing the extensive use of the smartphone in IoT and IoE applications in this chapter, we consider that the smartphone is the ultimate IoT and IoE device.

Author details

Mehdia Ajana El Khaddar[1]* and Mohammed Boulmalf[2]

*Address all correspondence to: mehdia.ajana@gmail.com

1 Ecole Nationale Supérieure de L'informatique et d'Analyse des Systèmes (ENSIAS), Rabat, Morocco

2 International University in Rabat (UIR), Rabat, Morocco

References

[1] Islam SMR, Kwak D, Kabir MDH, Hossain M, Kwak KS. The Internet of Things for health care: A comprehensive survey. IEEE Access. 2015;**3**:678-708. DOI: 10.1109/ACCESS.2015.2437951

[2] Rouse M. Internet of Everything (IoE) [Internet]. Available from: http://internetofthingsagenda.techtarget.com/definition/Internet-of-Everything-IoE [Accessed: 7 March 2017]

[3] Mukherjee P. The Smartphone and the Internet of Things [Internet]. 2015. Available from: http://praxis.ac.in/the-smartphone-and-the-internet-of-things/ [Accessed: 7 March 2017]

[4] Fendelman A. How are Cell Phones Different From Smartphones? Is a Cell Phone the Same As a Smartphone? [Internet]. 2017. Available from: https://www.lifewire.com/cell-phones-vs-smartphones-577507 [Accessed: 7 March 2017]

[5] Zheng P, Ni LM. Spotlight: The rise of the smart phone. IEEE Computer Society. 2006;**7**(3). DOI: 10.1109/MDSO.2006.22

[6] Romero J. Smartphones: The Pocketable PC: Is Your Phone Smarter Than a Fifth Grader? [Internet]. 2010. Available from: http://spectrum.ieee.org/telecom/wireless/smartphones-the-pocketable-pc [Accessed: 7 March 2017]

[7] Madakam S, Ramaswamy R, Tripathi S. Internet of Things (IoT): A literature review. Journal of Computer and Communications. 2015;**3**:164-173. DOI: 10.4236/jcc.2015.35021

[8] Vongsingthong S, Smanchat S. Internet of things: A review of applications & technologies. Suranaree Journal of Science & Technology. 2014;**21**(4): 359-374

[9] Vongsingthong S, Smanchat S. Internet of Things: A Review of Applications & Technologies [Internet]. 2014. Available from: http://www.thaiscience.info/journals/Article/SJST/10966646.pdf [Accessed: 7 March 2017]

[10] Ajana M E, Boulmalf M, Harroud H, Elkoutbi M. RFID Middleware Design and Architecture. In: Dr. Cristina Turcu, editor. Designing and Deploying RFID Applications. Rijeka, Croatia: InTech; 2011. DOI: 10.5772/16917

[11] Floerkemeier C, Lampe M. RFID middleware design: Addressing application requirements and RFID constraints. In: Proceedings of Smart Objects Conference; October; Grenoble, France. 2005. pp. 219-224

[12] ZIH Corp. Electronic Product Code (EPC) RFID Technology [Internet]. 2017. Available from: https://www.zebra.com/us/en/resource-library/getting-started/rfid-printing-encoding/epc-rfid-technology.html [Accessed: 7 March 2017]

[13] Ishikawa T, Yumoto Y, Kurata M, Endo M, Kinoshita S, Hoshino F, Yagi S, Nomachi M. Applying Auto-ID to the Japanese Publication Business to Deliver Advanced Supply Chain Management, Innovative Retail Applications, and Convenient and Safe Reader Services [Internet]. 2003. Available from: http://cocoa.ethz.ch/downloads/2014/06/None_KEI-AUTOID-WH004.pdf [Accessed: 7 March 2017]

[14] Tutorials Point. Wi-Fi Wireless Communication [Internet]. 2015. Available from: https://www.tutorialspoint.com/wi-fi/wifi_tutorial.pdf [Accessed: 7 March 2017]

[15] Puy I. Bluetooth [Internet]. 2008. Available from: http://webuser.hs-furtwangen.de/~heindl/ebte-08ss-bluetooth-Ingo-Puy-Crespo.pdf [Accessed: 7 March 2017]

[16] Obaid T, Rashed H, Abou-Elnour A, Rehan M, Muhammad-Saleh M, Tarique M. Zigbee technology and its application in wireless home automation systems: A survey. International Journal of Computer Networks & Communications. 2014;**6**(4). DOI: 10.5121/ijcnc.2014.6411

[17] Burkard S. Near Field Communication in Smartphones [Internet]. Available from: https://www.snet.tu-berlin.de/fileadmin/fg220/courses/WS1112/snet-project/nfc-in-smartphones_burkard.pdf [Accessed: 7 March 2017]

[18] Sohraby K, Minoli D, Znati T. Wireless Sensor Networks: Technology, Protocols, and Applications. 1st ed. Hoboken, NJ: John Wiley & Sons; 2007. pp. 38-69

[19] Southwest Center for Microsystems Education (SCME) University of New Mexico. Introduction to Transducers, Sensors and Actuators [Internet]. 2011. Available from: http://engtech.weebly.com/uploads/5/1/0/6/5106995/more_on_transducers_sensors_actuators.pdf [Accessed: 7 March 2017]

[20] Khan R, Khan SU, Zaheer R, Khan S. Future internet: The Internet of Things architecture, possible applications and key challenges. In: Proceedings of 10th International Conference on Frontiers of Information Technology (FIT); 17-19 December; Islamabad, Pakistan. 2012. DOI: 10.1109/FIT.2012.53

[21] Yang Z, Peng Y, Yue Y, Wang X, Yang Y, Liu W. Study and application on the architecture and key technologies for IOT. In: Proceedings of the International Conference on Multimedia Technology (ICMT); 26-28 July 2011. pp.747-751

[22] Wu M, Lu T, Ling F, Sun J, Du H. Research on the architecture of Internet of Things. In: Proceedings of the 3rd International Conference on Advanced Computer Theory and Engineering (ICACTE), August 2010

[23] Atzori L, Iera A, Morabito G. The Internet of Things: A survey. Computer Networks. 2010;**54**:2787-2805

[24] Chaqfeh MA, Mohamed N. Challenges in middleware solutions for the Internet of Things. In: Proceedings of the International Conference on Collaboration Technologies and Systems (CTS); Denver. 2012. pp. 21-26

[25] Al-Fuqaha A, Guizani M, Mohammadi M, Aledhari M, Ayyash M. Internet of Things: A survey on enabling technologies, protocols, and applications. IEEE Communication Surveys & Tutorials. 2015;**17**(4):2347-2376. DOI: 10.1109/COMST.2015.2444095

[26] Ajana ME, Harroud H, Boulmalf M, Elkoutbi M, Habbani A. Emerging wireless technologies in e-health trends, challenges, and framework design issues. In: Proceedings of the International Conference on Multimedia Computing and Systems (ICMCS), 10-12 October 2012. pp. 440-445

[27] Ajana ME, Harroud H, Boulmalf M, Hamam H. FlexRFID: A flexible middleware for RFID applications development. In: Proceedings of the 6th International Wireless and Optical Networks Communications (WOCN) Conference; April; Cairo, Egypt. 2009.

[28] Ajana ME, Harroud H, Boulmalf M, Elkoutbi M. FlexRFID middleware in the supply chain: Strategic values and challenges. International Journal of Mobile Computing and Multimedia Communications. 2011;**3**(2):19-32. DOI: 10.4018/jmcmc.2011040102

[29] Ajana EM, Chraibi M, Harroud H, Boulmalf M, Elkoutbi M, Maach A. FlexRFID: A security and service control policy-based middleware for context-aware pervasive computing. International Journal of Advanced Research in Artificial Intelligence (IJARAI). 2014;**3**(10). DOI: 10.14569/IJARAI.2014.031004

[30] Banafa A. The Internet of Everything (IoE) [Internet]. 2016. Available from: https://www.bbvaopenmind.com/en/the-internet-of-everything-ioe/ [Accessed: 7 March 2017]

[31] Bajarin T. The Next Big Thing for Tech: The Internet of Everything [Internet]. 2014. Available from: http://time.com/539/the-next-big-thing-for-tech-the-internet-of-everything/ [Accessed: 7 March 2017]

[32] Bradley J, Loucks J, Macaulay J, Noronha A. Internet of Everything (IoE) Value Index: How Much Value Are Private-Sector Firms Capturing from IoE in 2013? [Internet]. 2013. Available from: http://internetofeverything.cisco.com/sites/default/files/docs/en/ioe-value-index_Whitepaper.pdf [Accessed: 7 March 2017]

[33] Holler J, Tsiatsis V, Mulligan C, Avesand S, Karnouskos S, Boyle D. From Machine-to-Machine to the Internet of Things: Introduction to a New Age of Intelligence. 1st ed. Cambridge, Massachusetts, United States: Academic Press; 2014. 352 p

[34] Li R, Luo H. Based on the Internet of Things the supermarket chain management information system development and safety stock research. In: Proceedings of the 2nd International Conference on Education Technology and Computer (ICETC); 22-24 June; Shanghai, China. 2010. pp. 368-371

[35] Wei Y. Design and realization of mobile information collection module in logistic Internet of Things unified information system. In: Proceedings of IEEE 3rd International Conference on Communication Software and Networks (ICCSN); 27-29 May; Xian, China. 2011. pp. 263-266

[36] El-Baz D, Bourgeois J, Saadi T, Bassi A. ALMA, a logistic mobile application based on Internet of Things. In: Proceedings of the IEEE International Conference on Green Computing and Communications (GreenCom) and IEEE Internet of Things and IEEE Cyber, Physical and Social Computing; 20-23 August; Beijing, China. 2013. pp. 355-358

[37] Soldatos J, Kefalakis N, Leontiadis N, Konstantinou N, Mitton N, Schmidt L, Dagher R, David M, Anggorojati B, Frattasi S, Prasad N, Donsez D, Pedraza G, Gama K. D-3.4b: Core aspire middleware infrastructure. 2010. Technical Report, Aspire FP7 Contract: ICT-215417-CP, 89 p

[38] Lin X. Logistic geographical information detecting unified information system based on Internet of Things. In: Proceedings of the 3rd International Conference on Communication Software and Networks (ICCSN); 27-29 May; Xian, China. 2011. pp. 303-307

[39] Government Office for Science. The Internet of Things: Making the Most of the Second Digital Revolution [Internet]. Available from: https://www.gov.uk/government/uploads/system/uploads/attachment_data/file/409774/14-1230-internet-of-things-review.pdf [Accessed: 7 March 2017]

[40] Ghose A, Biswas P, Bhaumik C, Sharma M, Pal A, Jha A. Road condition monitoring and alert application: Using in-vehicle Smartphone as internet-connected sensor. In: Proceedings of the IEEE 10th International Conference on Pervasive Computing and Communications (PERCOM Workshops); 19-23 March; Lugano, Switzerland. 2012. pp. 489-491

[41] Ren X, Jiang H, Wu Y, Yang X, Liu K. The Internet of Things in the license plate recognition technology application and design. In: Proceedings of the Second International Conference on Business Computing and Global Informatization (BCGIN); 12-14 October; Shanghai, China. 2012. pp. 969-972

[42] Haiying W, Long H, Xin Q, Hongbo W, Gechen L, Xianqing D. Simulation system of the performance of power battery for electrical vehicle based on Internet of Things. In: Proceedings of the 2012 International Conference on Measurement, Information and Control (MIC); 18-20 May; Harbin, China. 2012. pp. 681-684

[43] Shengguang L, Lin T, Yuanshuo Z, Rucai Z. Internet of Things for special materials transportation vehicles. In: Proceedings of the IEEE International Conference on Green Computing and Communications (GreenCom) and IEEE Internet of Things and IEEE Cyber, Physical and Social Computing; 20-23 August; Beijing, China. 2013. pp. 1891-1894

[44] Rohokale VM, Prasad NR, Prasad R. A cooperative Internet of Things (IoT) for rural healthcare monitoring and control. In: Proceedings of the 2nd International Conference on Wireless Communication, Vehicular Technology, Information Theory and Aerospace & Electronic Systems Technology (Wireless VITAE); 28 February–3 March; Chennai, India. 2011. pp. 1-6

[45] Jara AJ, Zamora MA, Skarmeta AF. Knowledge acquisition and management architecture for mobile and personal health environments based on the Internet of Things. In: Proceedings of the IEEE 11th International Conference on Trust, Security and Privacy in Computing and Communications (TrustCom); 25-27 June; Liverpool, UK. 2012. pp. 1811-1818

[46] Yang N, Zhao X, Zhang H. A noncontact health monitoring model based on the Internet of Things. In: Proceedings of the 8th International Conference on Natural Computation; 29-31 May; Chongqing, China. 2012. pp. 506-510

[47] Kulkarni A, Sathe S. Healthcare applications of the Internet of Things: A review. International Journal of Computer Science and Information Technologies (IJCSIT). 2014;**5**(5):6229-6232. DOI: 10.1.1.659.5696

[48] Xiao Y, Li W, Chen X-J, Liu B-Y, Wang L, Fang D-Y. An immune theory based health monitoring and risk evaluation of earthen sites with Internet of Things. In: Proceedings of the International Conference on Green Computing and Communications (GreenCom) and IEEE Internet of Things and IEEE Cyber, Physical and Social Computing; 20-23 August; Beijing, China. 2013. pp. 378-382

[49] Kyriazis D, Varvarigou T, Rossi A, White D, Cooper J. Sustainable smart city IoT applications: Heat and electricity management & eco-conscious cruise control for public transportation. In: Proceedings of the14th International Symposium and Workshops on a World of Wireless, Mobile and Multimedia Networks (WoWMoM)); 4-7 June; Madrid, Spain. 2013. pp. 1-5

[50] Farooq MU, Waseem M, Mazhar S, Khairi A, Kamal T. A review on Internet of Things (IoT). International Journal of Computer Applications. 2015;**113**(1). DOI: 10.5120/19787-1571

[51] Miraz MH, Ali M. A review on Internet of Things (loT), Internet of Everything (IoE) and Internet of Nano Things (IoNT). In: Proceedings of the ITA 2015; 8-11 September; Wrexham. 2015. pp. 219-224

[52] Zielinski JS. Internet of Everything (IoE) in Smart Grid [Internet]. 2015. Available from: https://www.researchgate.net/publication/275956622_Internet_of_Everything_IoE_in_Smart_Grid [Accessed: 7 March 2017]

[53] Mitchell S, Villa N, Stewart-Weeks M, Lange A. The Internet of Everything for Cities: Connecting People, Process, Data, and Things to Improve the 'Livability' of Cities and Communities [Internet]. 2013. Available from: http://www.cisco.com/c/dam/en_us/solutions/industries/docs/gov/everything-for-cities.pdf [Accessed: 7 March 2017]

[54] Machination. The Development of the Internet of Everything [Internet]. 2014. Available from: https://www.machnation.com/wp-content/uploads/2014/06/MachNation-Development-of-the-IoE.pdf

[55] Hausenblas M. Smart Phones and the Internet of Things [Internet]. 2014. Available from: https://www.mapr.com/blog/smart-phones-and-internet-things [Accessed: 7 March 2017]

[56] Wittmann E. The Internet of Things Is Here, and It Will Revolve Around the Smartphone [Internet]. 2015. Available from: http://memeburn.com/2015/12/the-internet-of-things-is-here-and-it-will-revolve-around-the-smartphone/ [Accessed: 7 March 2017]

[57] Tempest A. The Internet of Everything [Internet]. 2014. Available from: http://www.rymote.com/general/internet-everything/ [Accessed: 7 March 2017]

[58] Ullah K, Shah MA, Zhang S. Effective ways to use Internet of Things in the field of medical and smart health care. In: Proceedings of the Intelligent Systems Engineering (ICISE) Conference; 15-17 January; Islamabad, Pakistan. 2016. DOI: 10.1109/INTELSE.2016.7475151

[59] Conrad D. Three Reasons Carriers Are Building New Cell Networks for the Internet of Things [Internet]. 2016. Available from: https://techcrunch.com/2016/10/28/three-reasons-carriers-are-building-new-cell-networks-for-the-internet-of-things/ [Accessed: 7 March 2017]

[60] Nokia. LTE Evolution for IoT Connectivity [Internet]. 2017. Available from: http://resources.alcatel-lucent.com/asset/200178 [Accessed: 7 March 2017]

[61] Ericsson. Cellular Networks for Massive IoT [Internet]. 2016. Available from: https://www.ericsson.com/res/docs/whitepapers/wp_iot.pdf [Accessed: 7 March 2017]

[62] FTC Staff Report. Internet of Things: Privacy & Security in a Connected World [Internet]. 2015. Available from: https://www.ftc.gov/system/files/documents/reports/federal-trade-commission-staff-report-november-2013-workshop-entitled-internet-things-privacy/150127iotrpt.pdf [Accessed: 7 March 2017]

[63] The Internet of Things: An Introduction to Privacy Issues with a Focus on the Retail and Home Environments [Internet]. 2016. Available from: https://www.priv.gc.ca/en/opc-actions-and-decisions/research/explore-privacy-research/2016/iot_201602/

[64] Khan MH, Shah MA. Survey on security threats of smartphones in Internet of Things. In: Proceedings of the International Conference on Automation and Computing (ICAC); 7-8 September; University of Essex Wivenhoe Park Colchester, UK. 2016. DOI: 10.1109/IConAC.2016.7604979

[65] Haley K. Mobile Apps and IoT Devices Are an Overlooked Security Risk by Consumers—And that's a Problem [Internet]. 2016. Available from: https://uk.norton.com/norton-blog/2016/02/mobile_apps_and_iot.html [Accessed: 7 March 2017]

Chapter 8

Positioning Techniques with Smartphone Technology: Performances and Methodologies in Outdoor and Indoor Scenarios

Paolo Dabove, Vincenzo Di Pietra and
Andrea Maria Lingua

Additional information is available at the end of the chapter

http://dx.doi.org/10.5772/intechopen.69679

Abstract

Smartphone technology is widespread both in the academy and in the commercial world. Almost every people have today a smartphone in their pocket, that are not only used to call other people but also to share their location on social networks or to plan activities. Today with a smartphone we can compute our position using the sensors settled inside the device that may also include accelerometers, gyroscopes and magnetometers, teslameter, proximity sensors, barometer, and GPS/GNSS chipset. In this chapter we want to analyze the state-of-the-art of the positioning with smartphone technology, considering both outdoor and indoor scenarios. Particular attention will be paid to this last situation, where the accuracy can be improved fusing information coming from more than one sensor. In particular, we will investigate an innovative method of image recognition based (IRB) technology, particularly useful in GNSS denied environment, taking into account the two main problems that arise when the IRB positioning methods are considered: the first one is the optimization of the battery, that implies the minimization of the frame rate, and secondly the latencies due to image processing for visual search solutions, required by the size of the database with the 3D environment images.

Keywords: positioning techniques, image navigation, GNSS, sensors, GPS

1. Introduction

Nowadays, thanks to new technologies, the information about our position is available in almost every moment and almost everywhere thanks to mobile devices, such as smartphones or tablets. These devices may include many sensors, such as global positioning system (GPS)/ global navigation satellite system (GNSS) chipset, inertial measurement unit (IMU) platforms, barometer, altimeter, cameras, etc., that empower customers to plan their activities (e.g., to know the time that it is necessary to wait a train) or to share their location on social networks (e.g., Facebook) [13]. With these kind of sensors and to the rise of new positioning techniques, it is possible to obtain the position both in outdoor and in indoor scenarios. In the first case, GPS/GNSS are the most useful sensors for obtaining a fast position even if there are some problems, especially in harsh environment, due to multipath or satellite obstructions. In the second case, these sensors became useless because no satellites are visible: so, it is possible to perform positioning thanks to other sensors, such as IMUs and cameras, considering other techniques such as the image recognition-based (IRB) or the pedestrian dead reckoning (PDR) technology.

In this chapter, we will investigate the positioning performances and methodologies in outdoor and indoor scenarios considering smartphone technology. In particular, the goal of this work is to analyze the state-of-the-art of the precisions and accuracies that can be achieved with these instruments for positioning and navigation purposes, in both scenarios.

In Section 2, the analysis of the most common sensors installed into smartphones is given as well as the methodology for the determination of the smartphone's reference system. We will discuss about the GNSS chipset (Section 2.1) available today and the positioning accuracy obtainable today with these sensors and with INS platforms (Section 2.2). Moreover a short description of cameras installed today into smartphones is done, in order to perform positioning using also images (Section 2.3).

Subsequently, in Section 3 a description of the positioning techniques obtainable today with smartphones is made and some practical examples are provided: the tests performed and the results obtained are presented, focusing the attention on outdoor (Section 3.1) and indoor (Section 3.2) scenarios.

Finally, some conclusions will be drawn in Section 4.

2. Sensors on smartphones

Many sensors are available today on smartphones: most of them are related to internal applications (proximity sensor, light sensors, etc.) while other ones (e.g., GNSS, INS, and cameras) allow to obtain a positioning. One of the biggest problems is represented by the operating system (OS) installed inside the smartphone: each OS has different ways to manage data that comes from internal sensors, not to mention the use of these data made by the apps.

While sensor availability varies from device to device, it can also vary between iOS and Android versions. The biggest changes were made in this last OS, due to several platform releases: in Android 1.5 (API Level 3) many sensors have been introduced even if some of them were not employed and not accessible before Android 2.3 (API Level 9). Similarly, in Android 2.3 (API Level 9) and Android 4.0 (API Level 14) some other sensors have been introduced and some others have been removed and replaced by newer ones.

Figure 1 shows the availability of each sensor on a platform-by-platform basis, considering the only four platforms that involved sensor changes.

In this chapter, we focus the attention only on sensors useful for positioning: GNSS and INS chipset and cameras for images. Hereinafter, a brief description of these sensors is provided.

2.1. GPS/GNSS chipsets

GPS/GNSS chipset is the most widespread sensor installed inside smartphones. There are many chipsets today available on the market, and very often each manufacturer installs few different versions of the same GNSS brand [2]. For example, Apple installs chips provided by Broadcom Corporation, while Samsung smartphones with Android OS have installed u-blox AG chipsets. Since 2016, no GNSS raw data acquired by mobile platform such as smartphones or tables were available, but starting from 2016 it has been possible also to extract

Sensor	Android 4.0 (API -- 14)	Android 2.3 (API -- 9)	Android 2.2 (API -- 8)	Android 1.5 (API -- 3)
TYPE_ACCELEROMETER	Yes	Yes	Yes	Yes
TYPE_AMBIENT_TEMPERATURE	Yes	n/a	n/a	n/a
TYPE_GRAVITY	Yes	Yes	n/a	n/a
TYPE_GYROSCOPE	Yes	Yes	n/a[1]	n/a[1]
TYPE_LIGHT	Yes	Yes	Yes	Yes
TYPE_LINEAR_ACCELERATION	Yes	Yes	n/a	n/a
TYPE_MAGNETIC_FIELD	Yes	Yes	Yes	Yes
TYPE_ORIENTATION	Yes[2]	Yes[2]	Yes[2]	Yes
TYPE_PRESSURE	Yes	Yes	n/a[1]	n/a[1]
TYPE_PROXIMITY	Yes	Yes	Yes	Yes
TYPE_RELATIVE_HUMIDITY	Yes	n/a	n/a	n/a
TYPE_ROTATION_VECTOR	Yes	Yes	n/a	n/a
TYPE_TEMPERATURE	Yes[2]	Yes	Yes	Yes

Figure 1. Availability of each sensor in different Android systems (available at: http://rowdysites.msudenver.edu/~gordona/cs390-mobile/lects/summer14_day07-touch+sensing/summer14_day07-sensors/summer14_day07-sensors.html).

pseudoranges and carrier-phase measurements from smartphones with Android 7.0 OS. The announcement came from Google during the I/O 2016, the 3-day developer conference which took place from 18 to 20 May. It is a very strong innovation, destined to bring a revolution in the field of survey and geo-localization: with these kinds of sensors, accuracies of few centimeters will be obtainable even with mobile devices. Despite that, this kind of possibility will not be analyzed in this chapter.

2.2. INS

Inertial measurement unit platforms are increasingly being used integrated either with other instruments, typically GNSS, odometers, and magnetometers, or with storage units [34]. They then form an inertial navigation system (INS).

In general, INS instruments are comprised of three accelerometers, three gyroscopes, and three magnetometers. The characteristics of these sensors are briefly described.

Accelerometers are instruments that measure acceleration (the rate of change in velocity), and help the phone distinguish up from down.

All accelerometers have two fundamental parts:

1. A housing attachment to the object whose acceleration we want to measure.
2. A mass that, while tethered to the housing, can still move.

For example, let us assume a spring and a heavy ball. If you move the housing up, the ball lags behind stretching the spring. If we measure how much that spring stretches, we can calculate the force of gravity.

Gyroscopes are sensors that can provide orientation information as well, but with greater precision. Thanks to this particular sensor, Android's Photo Sphere camera feature can tell how much a phone has been rotated and in which direction.

The digital compass is normally based on a sensor called magnetometer, which provides a simple orientation in relation to the Earth's magnetic field. Consequently, every smartphone knows where is North so it can autorotate the digital maps depending on its physical orientation.

2.3. Cameras

Like GPS/GNSS chipset, camera sensors are mandatory components for any kind of mass-market communication device and in particular for smartphones. CMOS image sensor (CIS) has always been one of the most important features in our phone, so that it was able to move the market to a new category of smartphones, the so-called camera phones. Google Pixel, Apple iPhone 7 Plus, Samsung Galaxy S7, Huawei P9, and Sony Xperia X are some of the best camera phones according to the international mobile industry congress like the International Consumer Electronics Show (CES) 2017 in Las Vegas and the Mobile World Congress (MWC) 2017 in Barcelona.

The mobile image sensing is the more direct way for the commercial user to represent the reality, to share information, and to create social interactions. For these reasons, in the last years, the CIS market has put particular emphasis on the quality of the camera modules, reaching high resolution levels, and makes them usable as low cost tools for numerous applications. Jointly with the numerous embedded sensors, like gyroscope, accelerometer, proximity sensor, GPS receiver, and Wi-Fi connectivity, these new camera chipset have boost the effort in R&D for new kind of applications like 3D sensing technology, such as Google's Tango, automotive self-drive, drone product, and virtual and augmented reality. Numerous technological upgrades in chipset architecture, like backside illumination (BSI) and in-sensors setups, like the dual camera implementation, have moved the market in favor of company rather than another. Nowadays, production and technology leader of image chipsets is Sony, covering 35% of the entire market. Sony's sensors are mounted into numerous smartphones and tablets like the Samsung Galaxy S7, Huawei P9, Sony Xperia X, etc. After, there is Samsung (Samsung Galaxy S7, Lenovo Vibe Shot) and Omnivision (Huawei P8, Lenovo K3 Note) and, according to many CIS market research firms, all three together reach about the 70% of the world market.

CMOS stands for complementary metal-oxide semiconductor, and it uses the same manufacturing technologies of CCDs sensors, the dominated technology till now, but needs much less power and the production is less expensive. The main advantages of CMOS imagers are that they are compatible with mainstream silicon chip technology and this allows on-chip processing and consequently the miniaturization. With the technological development on the semiconductor industry, the gap between CCD and CMOS has narrowed and the quality of the obtained image is competitive.

A typical CMOS is an integrated circuit with an array of pixel sensors and has the following main part:

- Micro lenses
- Color filter
- Pixel array
- ADC (analog to digital converter)
- Digital controller

Looking at the best camera phone of 2017, it is possible to state the specification of the camera sensors and make an overview on the best characteristics. **Table 1** resumes some smartphone camera specifications.

More details about cameras are available at the following hyperlinks:

- https://en.wikipedia.org/wiki/Exmor
- http://www.chipworks.com/sites/default/files/Apple_iPhone_6s_A1688_Smartphone_Chipworks_Teardown_Report_BPT-1509-801_with_Commentary.pdf
- http://www.gsmarena.com/samsung_galaxy_s7_camera_sensors_compared_sony_vs_samsung-news-17183.php

CMOS imaging sensor characteristics							
Smartphone	**Sensor name**	**Size (diagonal) [mm]**	d_{pix} **[μm]**	**CMOS technology**	**Sensor dimensions [mm × mm]**	**Image dimension [pix × pix]**	**MP**
Google Pixel/ BlackBerry Key One	Sony Exmor RS IMX378	7.81	1.55	BSI CMOS	6.25 × 4.69	4032 × 3024	12.2
Apple iPhone 6S	Sony Exmor RS IMX315	6.15	1.22	BSI CMOS	4.92 × 3.70	4032 × 3024	12.2
Apple iPhone 7	Sony Exmor RS IMX*	6.15	n/a	n/a	n/a	n/a	12
Apple iPhone 7 Plus	Sony Exmor RS IMX*	5	n/a	n/a	n/a	n/a	12
Samsung Galaxy S6, S6 Edge(+)	Sony Exmor RS IMX240	6.83	1.12	BSI CMOS	5.95 × 3.35	5312 × 2988	15.9
Samsung Galaxy S7, S7 Edge	Sony Exmor RS IMX260	7.06	1.4	BSI CMOS	5.64 × 4.23	4032 × 3024	12.2
	Samsung Isocell S5K2L1	7.06	1.4	ISOCELL	5.64 × 4.23	4033 × 3024	12.2
Huawei P9	Sony Exmor RS IMX286	6.2	1.25	BSI CMOS	4.96 × 3.72	3968 × 2976	11.8
OnePlus 3T/LGV20/ Huawei Mate 8/Asus Zenfone 3	Sony Exmor RS IMX298	6.4	1.12	BSI CMOS	5.16 × 3.87	4608 × 3456	15.9
Sony Xperia XZ	Sony Exmor RS IMX300	7.87	1.08	BSI CMOS	6.46 × 4.47	5984 × 4140	24.8
					5.96 × 4.47	5520 × 4140 (4:3 mode)	22.8
					6.46 × 3.64	5984 × 3366 (16:9 mode)	20.1
Sony Xperia XZ Premium (coming soon)	Sony Exmor RS IMX400	7.73	1.22	BSI CMOS	6.17 × 4.63	5056 × 3792	19.2
LG G4 e G5	Sony Exmor RS IMX234	6.83	1.12	BSI CMOS	5.95 × 3.35	5312 × 2988	15.9

Table 1. CMOS image sensor characteristics for commercial camera phones.

http://dx.doi.org/10.5772/intechopen.69679

- https://en.wikipedia.org/wiki/Samsung_CMOS
- https://www.androidheadlines.com/2016/04/huawei-p9-p9-plus-feature-sonys-12mp-imx286-sensors.html

3. Positioning with smartphones: outdoor and indoor scenarios

When outdoor scenarios are considered, smartphone technology can provide positions with a quite good level of accuracy, using the assisted GPS (A-GPS) system. Despite that, it is possible that the received GPS/GNSS signal is too noisy or not available at all, for example, if the user is in urban canyons or inside buildings: in these cases GNSS positioning is not possible.

Starting from that, many researchers have been investigating alternative solutions that consider different sensors (such as INS and images) and other technologies (e.g., Wi-Fi, pedestrian tracking system, Bluetooth) in order to improve position accuracy and availability. A brief overview about accuracies obtainable today with a generic smartphone (chosen as representative) is made in the following subsections.

3.1. Outdoor scenarios

3.1.1. GPS/GNSS only

As said in Section 2.1, starting from the end of 2016, it is possible to acquire raw GNSS measurements from smartphones: the main problem is that only Android Nougat OS allows to extract these information. Thus, in this section the attention is focused only on internal solutions provided by software installed on smartphones. In order to analyze the precision obtainable today with GNSS internal chipset, some tests were performed.

The tests took place in the same places described in [11] considering two different scenarios: an open outdoor area to represent "ideal" conditions (**Figure 2**, left) and another area (one of the courts in Politecnico di Torino campus) with characteristics of urban canyon (**Figure 2**, right). The line in **Figure 2** - right shows a particular track where it is possible to find an area with a limited satellite visibility (similar to urban canyon conditions) and many windows that create multipath due to their high reflectivity.

Dynamic tests were performed in these areas by walking along the same path with the smartphones mounted on a special "two-hands" support as shown in **Figure 3**. The entire data collection system includes:

- a smartphone (a)
- a 360-degree retro-reflector (d)
- and allows to install also an external IMU platform (b) and an external GNSS antenna (c).

GNSS data positions were recorded during the surveys considering a one-second sample rate, using a dedicated app that stores the National Marine Electronics Association (NMEA)

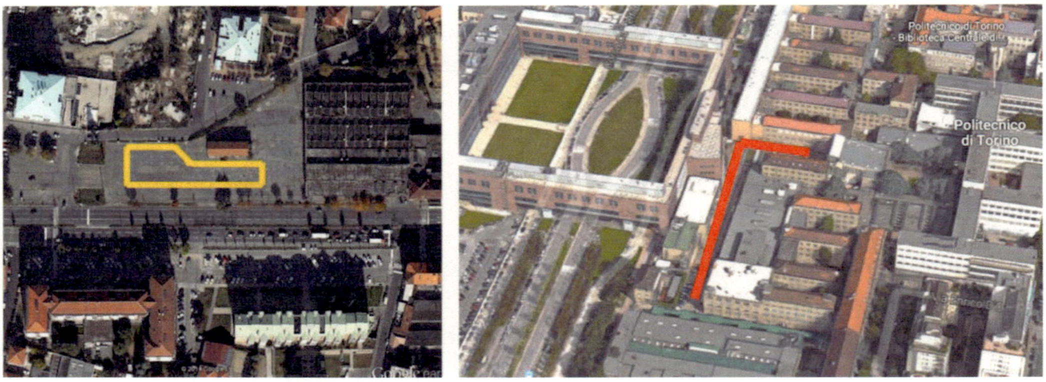

Figure 2. Test site and track: an open sky area (left) and an urban canyon (right).

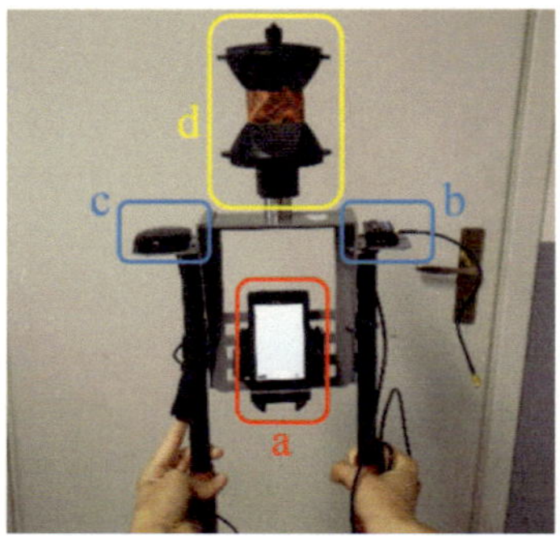

Figure 3. The two-hands support system developed at Politecnico di Torino.

GGA messages in an ASCII file. All results were compared with a "ground-truth" obtained through the continuous tracking of the smartphone position with a total station, thanks to the retro-reflector installed on the "two-hands" support. In this way, a millimeter accuracy was obtained, considering and estimating the level-arm offset between the instruments.

The NMEA sentences were analyzed and compared with the reference trajectories using software written in MATLAB.

The horizontal positioning errors of the representative receiver are shown in **Figure 4** for the urban canyon environment.

In order to have a more complete analysis from a statistical point of view, the most significant statistical parameters have been summarized in **Table 2** for the urban canyon and open area test locations.

As expected, then, it is generally possible to affirm that some environmental characteristics, such as obstacles, multipath effects coupled with the number of trackable satellites, play a crucial role in the accuracy determination of the smartphone positioning.

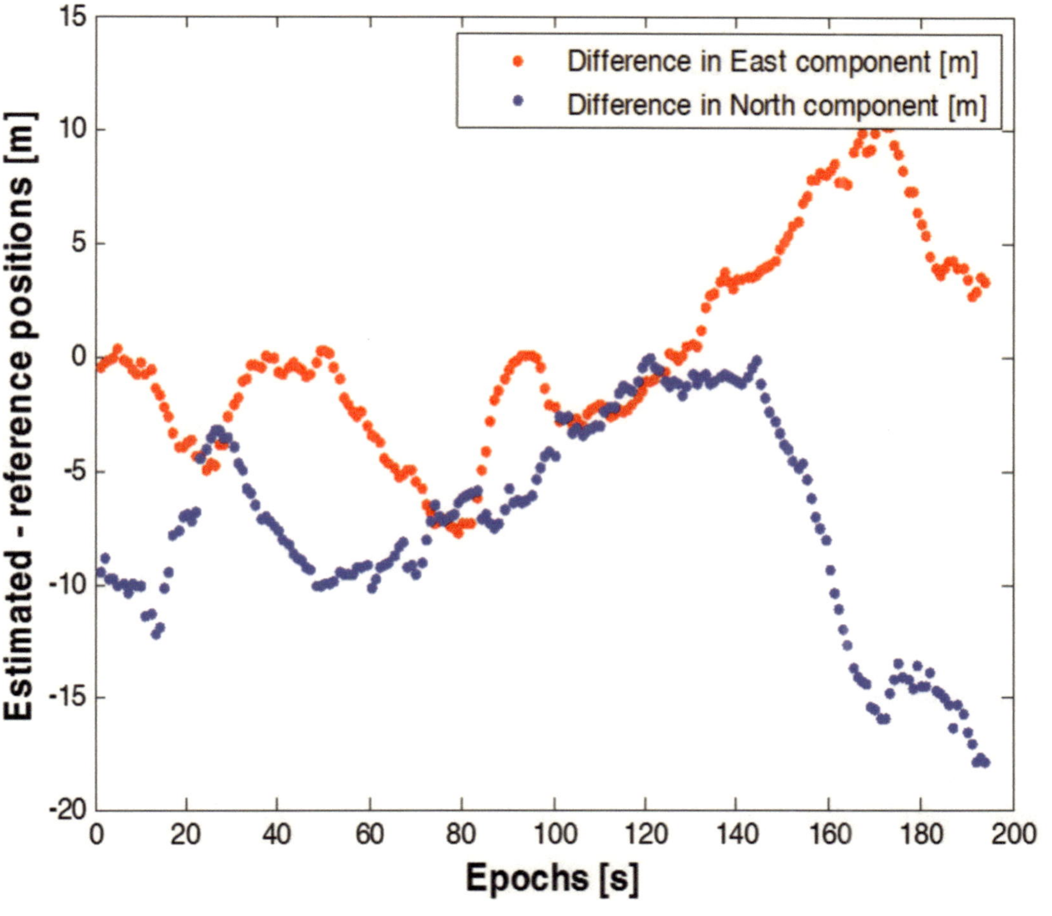

Figure 4. 2D performances of internal GPS sensor.

Smartphone	Mean (m)			Standard deviation (m)		
	E	N	H	E	N	H
Urban canyon environment	0.4	−7.3	−2.1	4.5	4.7	5.0
Open area environment	−0.5	1.6	−1.9	2.6	2.5	4.5

Table 2. Error statistics in urban canyon and open area environments.

However, precision and accuracy improvements could be increased by computing a differential positioning solution, considering the raw measurements obtainable from internal sensors.

3.1.2. GNSS + INS

Very often in bibliography it is possible to find two different methods for GNSS + INS positioning: the loosely (LC) and the tightly (TC) coupled approaches(**Figure 5**).

In the first case (LC), the software integrates acceleration and angular velocities and updates all the state parameters. These include the positions and angular assets, but also the instrumental biases, using GNSS positions and IMU measurements.

In the latter one (TC method), the input parameters are the same, but both the GNSS (pseudoranges, carrier-phases, Doppler) and IMU observations enter into the extended Kalman filter, each with its own rate and precision, associated with their new biases [8], in order to provide an unique solution.

Although from a computing point of view, this is a heavier process and it takes into account the use even of one or a few visible GNSS satellites, which is a typical situation of urban canyons [13]

It is important to underline that the only LC method is available today because this approach does not require the raw GNSS measurements, while for the TC is fundamental to have these observations.

Starting from this, in this subsection a brief analysis of results obtainable with GPS + INS instruments installed on smartphones is made, following the LC approach.

The tests were carried out in our campus in the same two different test sites described in the previous subsection, considering the same special support created in the Geomatics Lab at the Politecnico di Torino.

Considering the Inertial Explorer® software for postprocessing all data acquired on the field, it is possible to have an horizontal error loop equal to 4.21 m and a vertical error loop equal to 3.73 m, considering a 3-min session duration. Obviously, the results are slightly different if different smartphones are considered but it is possible t o affirm that these values are representative for the technology available today.

3.2. Indoor scenarios

The spread of smartphone devices with different embedded sensors, increased computational power and advance connectivity features, has led to the introduction into the market

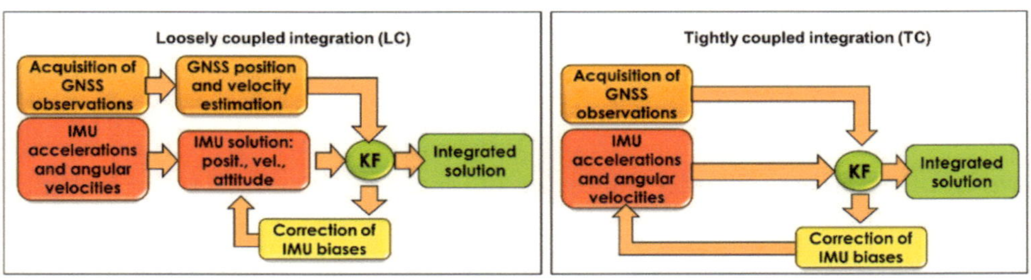

Figure 5. GNSS + INS processing approaches.

of numerous application services, based on the awareness of the user position, which provides information and assistance for navigation in the environment, pose estimation, tracking, and any kind of service related to the spatial context. Many location-based services (LBS) are implemented as information systems that use as prior information the position of a mobile device [22]. The number of companies deploying LBS solution for commercial purpose reveals that location-based solutions are finally meeting markets' needs and soon will be implemented on mass-market application. The principal fields of application are medical care [15], ambient assisted living [40], environmental monitoring [33], transportation [38], and marketing [1] etc.

The major part of these services requires accurate localization for people, instruments, vehicles, animals, and assets. As it is well known, the GNSS positioning provides good accuracies only in open-sky environments. Contrariwise, when an indoor space or in an urban canyon is considered, the GNSS positioning in not possible and it is mandatory to overcome this issue considering different techniques and sensors. In recent years, some indoor location-based services (LBSs) have been developed integrating different technologies and measurements [22], such as cameras [27], infrared (Kinect), ultrasound [20], WLAN/Wi-Fi [6], RFID [23], mobile communication [10], and so forth are examples of the technologies that the scientific community has put at the service of indoor locations. Despite the ample panorama of solutions, mass market applications for indoor positioning require the use of embedded sensors in commercial smartphone without supplementary physical components. For this reason, major modification to the devices is forbidden and the type of technology usable in these applications is reduced. Ref. [36] has made a summary on the user requirements for mass-market localization systems that is reported in **Table 3**.

All these indoor positioning systems have pros and cons that make them more useful in specific scenarios, compared to other options. One of the most useful but complex localization method is the inertial navigation system. This system is based on dead reckoning, which computes locations employing inertial measurements units installed inside the smartphone as accelerometers and gyroscope. The main advantages of a system using IMU (INS) is that nowadays, every kind of mobile device have it already implemented inside and no external

Criteria	Criteria description	Value
Horizontal accuracy	2D position for the detection of a shelf in a supermarket	1 m
Vertical accuracy	Selection of the correct floor and visualization	Floor detection
Update rate	Minimum for navigation	1 Hz
Latency	Delay with which position is available to the user	None
TTFF	Time-To-First-Fix, latency after switching on the device	Without delay
Privacy	Maintenance of the user privacy	According to user-set policy

Table 3. Summary of requirements for mass-marked localization according to Wirola et al. [36].

infrastructure is required. Moreover, with the inertial systems, the only input information that is needed is the staring position. Without any other external information required, this technology is not affected by adverse weather conditions or by security vulnerability or jamming problems. However, these systems suffer from integration drift, making errors accumulate and therefore must be corrected by some other system. The LBSs based on the camera sensor have strong advantages and do not need to install any network of chipsets in the environment. All the primary sensors are already installed in the user device. In this case, the system could be considered low cost. Moreover, the positioning accuracy with these systems is usually more accurate in comparison to other systems. Furthermore, most of these systems based on triangulation, cannot determinate the orientation of the user, with important limitations to support many useful applications like augmented reality.

It is evident now that providing a reliable and stable position information in a complex and changing environment is a very challenging task. Sensor fusion may be an option to combine advantages of two or more different techniques (e.g., angle of arrival (AoA), time of flight (ToF), received signal straight indication (RSSI)), and technologies (e.g., GPS, Wi-Fi, Bluetooth, camera sensors, ultrasound) and minimize the limitation. Some methods and technologies are ideal candidate to support or complete other navigation or localization systems in a multimodal approach in order to obtain an accuracy and reliability in the location information superior to that obtainable by each technique, technology, or system parameter without the use of diversity. Multimode solutions employing different sensor would not be feasible for low-end handsets unable to connect to more than one technology or without the hardware enhancements required to apply different techniques. For these reasons, the positioning solution with smartphone technologies exploits the already embedded sensors: INS, CMOS image sensor, and Wi-Fi.

In this chapter, we will focus on solution of image recognition-based (IRB) technology that uses CIS as the main sensor. In particular, after a general overview on existing methods, we will investigate an innovative method of IRB location based on the image retrieval of real-time acquired smartphone pictures with the corresponding synthetically generated 3D image or RGBD image extracted by a database. Then we will evaluate the integration of INS for a multimodal solution of the previous method for indoor navigation and finally some consideration on system using Wi-Fi technology as the main positioning technology.

3.2.1. Cameras + INS

Indoor positioning and navigation by optical sensors is becoming one of the dominant techniques, able to cover a large number of fields of application at all levels of accuracy. The success of these techniques is due to the improvements and miniaturization of the CMOS sensors. Simultaneously, there has been an increase in the data transfer speed and smartphone computational capabilities, as well as a remarkable development in the field of image processing.

As seen before, the LBSs based on the camera sensor have strong advantages. First, these systems do not need to install any network of chipsets in the environment as the primary sensor (CIS) is already installed in the user device. This allows to develop a low cost service without design and implementation of onsite network. Moreover, the positioning accuracy with these systems is usually more accurate in comparison to other systems. In industrial process, for example, computer vision systems based on object detection algorithms are used in production line to track object and check the quality. These kinds of systems have accuracy around few millimeters. Of course, applications of image-based positioning with smartphones cannot reach these levels of accuracy but can perfectly match the requirements for navigation purpose.

There are many previous research studies on indoor image-based localization that pursue different goals and use different methods and technologies also in the function of the field of interest of the research groups. There are visual odometry approaches [19], simultaneous and location mapping (SLAM) [24], structure from motion, or investigating semantic features [29]. Some interesting work exploits the computer vision algorithm and in particular the neural network and transfer learning for visual indoor positioning and classification [35]. Some use RGB-D images to perform object recognition [25]. On the use of a smartphone as a navigation device, some interesting research can be found in [27, 39].

As seen in bibliography, there are many LBS based on images, whose accuracies and coverage area is function of the application. Some accuracy ranges may be useful for applications in very large indoor spaces like museums or fairs, and others may require accuracies at sub-room level, for example, in the field of logistics and optimization. When trying to make indoor positioning and navigation in more complex spaces with task of "search and rescue" or in construction sites, the coverage area decreases and higher accuracies are required.

A possible solution, considering all sensors which are installed into the smartphone device, is the image recognition-based approach, where the localization of our device is based on photogrammetric principle [16]. Image recognition-based (IRB) positioning is a good technology for smartphone indoor localization. The aim of these procedures is to match a user-generated query image, via a mobile device, against an existing image database with position information [41].

Some test has been carried out in our campus, following the methodologies presented in [9, 28]. The use of IRB positioning in mobile applications is characterized by the availability of a single camera; under this constraint, in order to estimate the camera parameters (position and orientation), a prior knowledge of 3D environment has to be available, in the form of a database of images with associated spatial information. A Terrestrial LiDAR (Light Detection and Ranging) Survey (TLS) with an associated camera can be executed to acquire the 3D model of the environment used to generate the images database (RGB-D images) . Once the retrieval of the reference image is completed, it is possible to extract the 3D information of the selected features from the image to estimate the external parameters (position and attitude) of the query image according to the collinearity equations (**Figure 6**).

A priori information are necessary for these techniques, but nowadays, an accurate 3D model that could be always available for further upgrading and be usable for collateral tasks is obtainable, thanks to the integration of some geomatics techniques, such as photogrammetry, LiDAR, and mobile mapping systems.

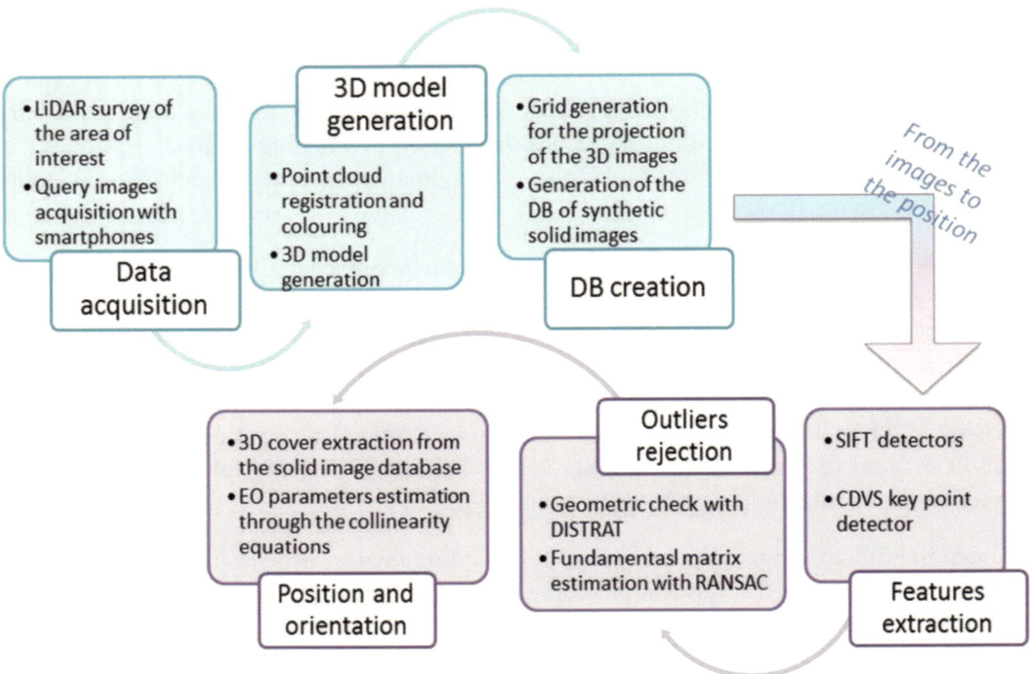

Figure 6. The IRBL procedure.

Table 4 summarizes the accuracy results in terms of discrepancies from ground truth and estimated values, for indoor trial in case of good level of similarity between the query image and the reference one extracted out of the database.

It is important to state that the entire procedure could be executed in real time on a commercial smartphone and could provide the device position in few seconds. This is true for one-spot positioning task, while, when it is needed to transpose the methodology for indoor navigation, it is necessary to take into account three fundamental problems: the energy consumption, the latency of the image processing, and the Internet data consumption. Acquiring images at a given frame rate for navigation applications is a procedure that requires high wastage of energy, with a consequent problem of battery optimization. Furthermore, as each query image has to be sent to a server for image retrieval procedure, a certain amount of Internet traffic is needed. Finally, the rates of positioning information are subordinated to the latency of the entire IRB methodology.

To overcome the reduction of the frame rate and latencies compensation, inertial (INS) platforms built with MEMS (micro electro-mechanical systems) technology can be integrated in the IRB positioning [12]. Fusing IRB position and attitude measurement with INS measurements, accelerations, and angular velocity measurements are integrated to provide real-time relative position and relative attitude information, while inner INS variables (velocity at the starting point, accelerometer biases, and gyro drift) are estimated using absolute IRB positioning inputs (position and attitude).

http://dx.doi.org/10.5772/intechopen.69679

Param.	ΔX [m]	ΔY [m]	ΔZ [m]	$\Delta\omega$ [rad]	$\Delta\phi$ [rad]	Δk [rad]
|Max|	0.164	0.149	0.063	0.4646	0.5288	0.2396
Mean	0.018	0.010	0.015	0.0975	0.0544	0.0573
Dev. St.	0.084	0.086	0.020	0.1850	0.1689	0.1022

Table 4. Accuracy results in indoor trial for position (ΔX, ΔY, ΔZ) and attitude ($\Delta\omega$, $\Delta\varphi$, Δk).

When MEMS technology is used together with IRB positioning, it is important to analyze the precisions and accuracies obtainable. The procedure was tested in our campus walking in a predefined path using two different smartphones (a) mounted on a special support, as described in **Figure 3**.

The procedure starts with the analysis of the raw data of inertial sensors (acceleration, angular velocity, and magnitude of magnetic field), directly registered from the smartphone. It is necessary to filter these data for estimating and removing the noise. After that, it is possible to use INS raw data in real time for positioning purposes considering a Kalman filter approach in order to reduce the number of frames that can be acquired for geo-localization. This means that it is possible to extend the time interval between two images from 2 s up to 5 s, depending on the requested accuracy.

In **Table 5**, we see the positioning results in terms of accuracies, considering an IBN approach. Considering an interval of 1 s between images, the mean planimetric error was 21.3 cm at 67% of reliability, while at 95% this error was 37 cm.

When the positioning obtained with an interval of 2 s between the images is analyzed, the mean planimetric error increases to 61 cm at 67% and 1.49 m at 95%.

IBN allows to reduce to 50% the final residuals and increase the outages up to 90 s, even improving the quality of the estimated angles. At the moment, the IBN requires a server with high performance in order to obtain the solution and a well-defined images database (DB).

3.2.2. Wi-Fi et al.

Over the last decade, wireless location estimation has been an active field of research, becoming the most widespread approach for indoor localization in GNSS denied environment. A WLAN (Wireless Local Area Networks, IEEE 802.11 standard), otherwise known as Wi-Fi

	1 s			2 s			5 s		
	Mean	Percentile		Mean	Percentile		Mean	Percentile	
		67%	95%		67%	95%		67%	95%
E	0.130	0.148	0.353	0.387	0.485	0.960	1.673	2.221	4.158
N	0.130	0.141	0.412	0.380	0.409	1.162	1.574	1.677	3.952

Table 5. Results obtained with drift estimation coming from images.

(Wi-Fi is a trademark of the Wi-Fi Alliance), is a wireless network of devices that uses high frequency radio signal (2.4 GHz in ISM band) to transmit and receive data within a limited area. As the connection between nodes of the network maintains continuity, the communication is preserved even if one device is moving around in the limited area (50–100 m) [37]. This means that for these reasons, the WLAN technology could be used to estimate the location of a mobile device within this network. The positioning accuracy required to offer satisfactory LBSs is in the order of 1 m and a great effort is needed in R&D. The expansion of this field of research is expected to continue for years, beside numerous commercial applications, due to the fact that it is a low cost solution providing proper connectivity and high speed links. In fact, nowadays, the WLAN infrastructure is widespread in many indoor environments and it is already standardize for commercial smartphone communication.

Usually, an indoor environment is often complex, characterized by nonline-of-sight (NLOS) of target objects; in these situations, WLAN positioning technologies could be very helpful because they do not require the line of sight. Unfortunately, compared to IRBL procedure, WLAN positioning is affected by a large estimation error, proportional to the number, and position of nodes in the network. Others challenging issues are the power consumption and the signal attenuation.

Pros and cons of WLAN positioning are true in function of the techniques of positioning used. The most popular WLAN positioning method is based on the received signal straight indicator (RSSI) because it is easy to extract from any connected device in a Wi-Fi network [17]. The RSSI method is based on the received signal power and on the relation between the signal attenuation and distance of the nodes. Knowing the strength of the emitted signal, the strength of the received signal, is possible to calculate its attenuation and consequently the distance between the emitter and the receiver. With these techniques, it is possible to combine different strategies for positioning, like propagation modeling, fingerprinting, cell of origin, and multilateration [30]. To obtain a most precise localization, it is necessary to combine the technique of fingerprinting [37] that consists an *a priori* analysis to map the observed signal strength of fixed routers in every place of the indoor environment. With this data it is possible to generate a database (i.e., a radio map). The limitation of this method is the necessity of *a priori* information, an effort that means an increased workload and a well-spread router network. The propagation model differs from the fingerprinting model because it tries to determinate the RSSI map analytically instead of empirically. Of course, the major issues are related with the right description and modeling of the environmental effects (moving objects, signal attenuation, multipath) [5].

Another way to locate a device in a Wi-Fi positioning system is the cell of origin (CoO) method, with which the receiver position is made to coincide with the coordinate of the access point (AP) generating the highest RSSI value. Due to the spatial distribution of the APs in an indoor environment, this type of techniques is able to reach location with errors around 10–20 m [14].

Finally, multilateration methods, like time of arrival, time difference of arrival, angle of arrival, and so forth, are less common for WLAN positioning due to computational complexity of these kinds of measurements in mobile devices [26].

A literature review on WLAN systems for indoor positioning has been published by He et al. in 2016 [18]. There are many previous research studies on indoor Wi-Fi localization that pursue different goals and use different methods and technologies also in the function of the field of interest of the research groups. In particular, besides the numerous interesting works on positioning and navigation on self-made mobile devices with sensor integration, there are some researches exploiting the embedded sensors in COTS (Commercial On-The-Shelf) smartphone. Some interesting work exploits the integration of inertial sensor-based positioning with Wi-Fi capability of smartphones [7, 31]. For example, in [7] the authors propose a sensor fusion framework for combining Wi-Fi, pedestrian dead reckoning (PDR) and landmarks. The whole system runs on a smartphone and Android app is developed for real-time indoor localization and navigation. The established accuracy is 1 m. An interesting multimodal approach of Wi-Fi navigation is described in [21], where PDR carried out with only low cost sensors and Wi-Fi smartphones are issued in a cooperative positioning operation made by a certain number of participants. The size of the error becomes smaller when the number of participants rises (5 m for 50 devices). Some use GPS integration for cloud-based LBSs [3], while other researchers introduce sensors fusion between Wi-Fi and CIS for accurate indoor positioning [32] or for augmented reality navigation [1].

A comprehensive and complete view on indoor positioning systems implemented today, with its applications and obtainable positioning accuracies, is described in [18].

4. Conclusion

In this chapter, the authors have tried to describe the positioning performances and methodologies in outdoor and indoor scenarios considering smartphone technology. In particular, the state-of-the-art of smartphone technology regarding the precisions and accuracies that can be achieved with these instruments for positioning and navigation purposes has been analyzed, in both scenarios.

Even if in outdoor scenarios the obtainable accuracy is less than 5 m under open-sky conditions considering only the GNSS sensors, in indoor environment is possible to have accuracies of 10–50 cm if some sensors, such as INS, cameras, or Wi-Fi technology, are considered. Interesting results are obtained fusing IRB technique with MEMS technology: considering an interval of 2 s between images, the mean planimetric error is about 61 cm at 67% and 1.49 m at 95% of reliability. A possible interesting alternative for indoor positioning could be represented by the fusion of range camera and INS instruments.

This is not an exhaustive overview, also because the technology is evolving more quickly than the minds that are producing them: so this would be a starting point for future works regarding these instruments for positioning and mapping application.

Author details

Paolo Dabove*, Vincenzo Di Pietra and Andrea Maria Lingua

*Address all correspondence to: paolo.dabove@polito.it

Politecnico di Torino, Department of Environment, Land, and Infrastructure Engineering, Turin, Italy

References

[1] Alnabhan A, Tomaszewski B. INSAR: Indoor navigation system using augmented reality. In: ISA '14 Proceedings of the Sixth ACM SIGSPATIAL International Workshop on Indoor Spatial Awareness; November 04-04, 2014. Dallas/Fort Worth, Texas. New York, USA: ACM; 2014. pp. 36-43

[2] Bauer C. On the (In-)accuracy of GPS measures of smartphones: A study of running tracking applications. In: Proceedings of 11th International Conference on Advances in Mobile Computing & Multimedia (MoMM 2013); December 02-04, 2013. Vienna, Austria. New York, NY, USA: ACM; 2013. pp. 335-340

[3] Bisio I, Lavagetto F, Marchese M, Sciarrone A. GPS/HPS-and Wi-Fi fingerprint-based location recognition for check-in applications over smartphones in cloud-based LBSs. IEEE Transactions on Multimedia. 2013;**15**(4):858-869

[4] Bornaz L, Dequal S. A new concept: The solid image. In: Proceedings of the XIXth International Symposium, CIPA 2003: New Perspectives to Save Cultural Heritage; 30 September–4 October 2003. Antalya, Turkey.

[5] Bose A, Foh CH. A practical path loss model for indoor WiFi positioning enhancement. In: Information, Communications & Signal Processing, 2007 6th International Conference on; 10-13 Dec. 2007. Singapore: IEEE; 2007. pp. 1-5

[6] Bumgon K, Wonsun B, Kim YC. Indoor localization for Wi-Fi devices by cross-monitoring AP and weighted triangulation. In: Proceedings of the IEEE Consumer Communications and Networking Conference (CCNC); 9-12 January 2011. Las Vegas, NV, USA: IEEE; 2011

[7] Chen LH, Wu EHK, Jin MH, Chen GH. Intelligent fusion of Wi-Fi and inertial sensor-based positioning systems for indoor pedestrian navigation. IEEE Sensors Journal. 2014;**14**(11):4034-4042

[8] Chu CH, Chiang KW, Liao JK, Rau JY, Tseng YH, Chen JH, Chen JC. The performance of a tight INS/GNSS/photogrammetric integration scheme for land based MMS applications in GNSS denied environments. In: Shortis M, El-Sheimy N, editors. International Archives of the Photogrammetry, Remote Sensing and Spatial Information Sciences. Melbourne: Springer-Verlag; 2012. pp. 479-484

[9] Chiabrando F, Di Pietra V, Lingua A, Cho Y, Jeon J. An original application of image recognition based location in complex indoor environments. ISPRS International Journal of Geo-Information. 2017;**6**:56

[10] Cui X, Gulliver TA, Song H, Li J. Real-time positioning based on millimeter wave device to device communications. IEEE Access. 2016;**4**:5520-5530

[11] Dabove P. What are the actual performances of GNSS positioning using smartphone technology? Inside GNSS. 2014;**2014**:35-37

[12] Dabove P, Ghinamo G, Lingua AM. Inertial sensors for smartphones navigation. SpringerPlus. 2015;**4**(1):834

[13] Dabove P, Manzino A. Accurate Real-time GNSS positioning assisted by tablets: An innovative method for positioning and mapping. Geam. Geoingegneria Ambientale e Mineraria. 2016;**148**(2):17-22

[14] Farid Z, Nordin R, Ismail M. Recent advances in wireless indoor localization techniques and system. Journal of Computer Networks and Communications. 2013. pp.1-13

[15] Fisher JA. Indoor positioning and digital management: Emerging surveillance regimes in hospitals. In: Surveillance and Security: Technological Politics and Power in Everyday Life. New York, USA: Taylor & Francis; 2006. pp. 77-88

[16] Gupta A, Garg R, Kaminsky R. An Image-based positioning system. 2010. pp. 1-8 available at courses.cs.washington.edu

[17] Hatami A, Pahlavan K. A comparative performance evaluation of RSS-Based positioning algorithms used in WLAN networks. IEEE Wireless Communications and Networking Conference; 13-17 March 2005. New Orleans, LA, USA: IEEE; 2005

[18] He S, Chan SHG. Wi-Fi fingerprint-based indoor positioning: Recent advances and comparisons. IEEE Communications Surveys & Tutorials. 2016;**18**(1):466-490

[19] Huang AS, Bachrach A, Henry P, Krainin M, Maturana D, Fox D, Roy N. Visual odometry and mapping for autonomous flight using an RGB-D camera. In: Robotics Research. Berlin, Germany: Springer; 2017. pp. 235-252

[20] Ijaz F, Yang HK, Ahmad AW, Lee C. Indoor positioning: A review of indoor ultrasonic positioning systems. In: Proceedings of the 15 International Conference Advanced Communication Technology (ICACT); 27-30 January 2013. Pyeongchang, Korea: IEEE; 2013. pp. 1146-1150

[21] Kang W, Han Y. SmartPDR: Smartphone-based pedestrian dead reckoning for indoor localization. IEEE Sensors Journal. 2015;**15**(5):2906-2916

[22] Koyuncu H, Yang SH. A survey of indoor positioning and object locating systems. IJCSNS International Journal of Computer Science and Network Security. 2010;**10**(5):121-128

[23] Lau EEL, Chung WY. Enhanced RSSI-based real-time user location tracking system for indoor and outdoor environments. In: Proceedings of the International Conference on Convergence Information Technology; 21-23 November 2007; Gyeongju, Korea

[24] Levchev P, Krishnan MN, Yu C, Menke J, Zakhor A. Simultaneous fingerprinting and mapping for multimodal image and WiFi indoor positioning. In: Proceedings of the 2014 International Conference on Indoor Positioning and Indoor Navigation (IPIN); 27-30 October 2014; Busan, Korea. IEEE; 2014

[25] Li X, Fang M, Zhang JJ, Wu J. Learning coupled classifiers with RGB images for RGB-D object recognition. Pattern Recognition. 2017;**61**:433-446

[26] Li B, Salter J, Dempster AG, Rizos C. Indoor positioning techniques based on wireless LAN. First IEEE International Conference on Wireless Broadband and Ultra Wideband Communications; 13-16 March 2006; Sydney, Australia. 2006

[27] Liang JZ, Corso N, Turner E, Zakhor A. Image-based positioning of mobile devices in indoor environments. In: Multimodal Location Estimation of Videos and Images. Berlin, Germany: Springer; 2005. pp. 85-99

[28] Lingua A, Aicardi I, Ghinamo G, Corbi C, Francini G, Lepsoy S, Lovisolo P. The MPEG7 Visual Search Solution for image recognition based positioning using 3D models. In: Proceedings of the 27th International Technical Meeting of the Satellite Division of the Institute of Navigation (ION GNSS+ 2014); 8-12 September 2014; Tampa, FL, USA. 2014. pp. 2078-2088

[29] Liu L, Sisi Z. A semantic data model for indoor navigation. In: Proceedings of the Fourth ACM SIGSPATIAL International Workshop on Indoor Spatial Awareness; 6 November 2012; Redondo Beach, CA, USA. New York, NY, USA: ACM; 2012

[30] Mautz R. Indoor positioning technologies. Habilitation Thesis, ETH Zurich. 2012

[31] Panyov AA, Golovan AA, Smirnov AS. Indoor positioning using Wi-Fi fingerprinting pedestrian dead reckoning and aided INS. In: Inertial Sensors and Systems (ISISS); 2014 International Symposium on; 25-26 Feb. 2014. Laguna Beach, CA, USA: IEEE; 2014. pp. 1-2

[32] Papaioannou S, Wen H, Markham A, Trigoni N. Fusion of radio and camera sensor data for accurate indoor positioning. In: Mobile Ad Hoc and Sensor Systems (MASS), 2014 IEEE 11th International Conference on ; 28-30 Oct. 2014. Philadelphia, PA, USA: IEEE; 2014. pp. 109-117

[33] Tellez M; El-Tawab S, Heydari HM. Improving the security of wireless sensor networks in an IoT environmental monitoring system. In: Proceedings of the 2016 Systems and Information Engineering Design Symposium (SIEDS); 29 April 2016. Charlottesville, VA, USA: IEEE; 2016

[34] Titterton DH, Weston JL. Strapdown Inertial Navigation Technology. 2nd ed. VA, USA, Reston: The American Institute of Aeronautics and Astronautics; 2004

[35] Werner M, Hahn C, Schauer L. DeepMoVIPS: Visual indoor positioning using transfer learning. In: Proceedings of the 7th International Conference on Indoor Positioning and Indoor Navigation (IPIN); 5-7 October 2016. Madrid, Spain: IEEE; 2016

[36] Wirola L, Laine TA, Syrjärinne J. Mass-market requirements for indoor positioning and indoor navigation. In: Indoor Positioning and Indoor Navigation (IPIN), 2010 International Conference on; 15-17 Sept. 2010. Zurich, Switzerland: IEEE; 2010. pp. 1-7

[37] Wain Sy Au. RSS-based WLAN indoor positioning and tracking system using compressive sensing and its implementation on mobile devices [thesis]. University of Toronto; 2010; p. 136

[38] Xiao Z, Havyarimana V, Li T, Wang D. A nonlinear framework of delayed particle smoothing method for vehicle localization under non-Gaussian environment. Sensors. 2016;**16**:692

[39] Yang J, Xu R, Lv Z, Song H. Analysis of camera arrays applicable to the internet of things. Sensors. 2016;**16**:421

[40] Zetik R, Shen G, Thomä R. Evaluation of requirements for UWB localization systems in home-entertainment applications. In: Proceedings of the 2010 International Conference on Indoor Positioning and Indoor Navigation (IPIN); 15-17 September 2010. Zurich, Switzerland: IEEE; 2010

[41] Zhang Q, Niu X, Gong L, Zhang H, Shi C, Liu C, Wang J, Coleman M. Development and Evaluation of GNSS/INS Data Processing Software. China Satellite Navigation Conference (CSNC) 2013 Proceedings. Springer; 2013. pp. 685-696

Chapter 9

Where Is the Smartphone Leading the Health of Children?

Simlawo Kpatékana, Boumé Missoki Azanlédji,
Kanassoua Kokou,
Mihluedo-Agbolan Komlan Anani and
Bouame Kokou Tsolanyo

Additional information is available at the end of the chapter

http://dx.doi.org/10.5772/intechopen.70260

Abstract

Smartphones allow users to carry a mobile phone, games console, music player, camera, calendar, and Internet browser all in one small handheld device, with their limitations governed only by the types of applications downloaded onto them. They have become an indispensable part of the daily life. While smartphones have made life more convenient with their advantages, they have also brought many side effects especially on the health. This chapter crosses literature data on the side effects of smartphone in terms of health, especially in children. Nonetheless, it may affect people's psychology, behavior, and health especially those of children. A mobile phone battery when heated explodes as a bomb. Awareness should be raised on the dangers of smartphones for children as telephone has become a real life partner in everything. Telephony companies as well as parents should join their effort, and measures should be taken to protect children and teenagers to ensure their welfare as they use smartphones. It is not enough to say that the humanity is in permanent danger. It is necessary to prioritize the protection of health while we rejoice in these technological advances.

Keywords: smartphone, health, insecurity, trauma, children

1. Introduction

A smartphone is a category of mobile device that provides advanced capabilities software system that provides a standardized interface and platform for application developers. They allow users to carry a mobile phone, games console, music player, camera, and calendar and

Internet browser all in one small handheld device, with their limitations governed only by the types of applications downloaded onto them [1]. There are over 1.5 billion smartphone users around the world, and it was estimated that more than one billion smartphones will be sold in 2016 [2]. Smartphones have become an indispensable part of the daily life. While smartphones have made life more convenient with their advantages [3–5], they have also brought many side effects [2, 6] especially on the health. This chapter aims to focus on the benefaction and disadvantages of smartphone on the health.

This study concerns children less than 15 years of both sexes. We cross literature data on the side effects of smartphone in terms of health, especially in children. Genesis of smartphone, advantages of smartphones, and potential danger of their mismanagement, measures to limit smartphones' dangers would be studied. We will focus on differences between developed and developing countries.

2. Genesis of smartphone

A smartphone is a mobile phone that includes functions similar to those found on personal computers. They provide a one-stop solution for information management, mobile calls, emails sending, and Internet access [7].

It is of the large family of mobile phones and slightly larger than standard mobile telephones.

Mobile phone came into being by the end of 1940s with very limited functions. The very first mobile phone was heavy due to its size and is very expensive. Not everybody could afford it. There was no easy access to mobile phones and very few people could use them. Even if that phone was mobile, it was not easy to carry everywhere, and therefore it was used only in cars and some specific places. However, in 1982, the mobile phone of the first generation (1G) was created in the United States of America by the Bell Laboratories. It was a device with analogue voice. About a decade later, the mobile phone evolves to the second generation (2G). It became smaller, lighter, and cheaper. It functions on a Global System for Mobile Communication (GSM). Many people could afford a cellular then. It helped reducing the price of lines and promoted their integration into the business world since many conversations could be made simultaneously on a single channel. Communication became progressively easier and less expensive [8].

3. Advantages of smartphones

3.1. Mobile phone purchased for prestige and power

However, in the developing and underdeveloped countries, it was almost impossible to afford a mobile phone and only some highly positioned people could. It was something of prestige and luxury in those countries. Many of those who acquired mobile telephones would not buy it because they needed it for business purposes. It was just to show their purchasing power and their own level of riches. Many were suspicious vis-a-vis mobile phones. They had never seen it before and could not figure out its utility. Some even thought it was a trick from the white to control Africans. It took Africans quite a long time to get into the rhythm of the globalization of technology.

Making and receiving calls everywhere and at any time thanks to cell phones addressed some difficulties related to communication. Therefore, business started doing better for the difficulties are being suppressed progressively. Technology on the whole is evolving to a level that it has never reached before. Computers and network use are widespread; local area network (LAN) and wide area network (WAN) connections connected to desktops and laptops were set. Men and women who enjoy all the technology advancement would not content to these services and would want more. Everything is moving at a dazzling speed.

Science helped human beings to destroy the world at the time of WWI and WWII. Technology has come to help them rebuild it, so there is no way to waste time, it is now or never. Therefore, companies would address needs for data transmission (multimedia), leading to what is known as the third generation (3G) mobile phones with the development of the Universal Mobile Telecommunication System (UMS). They are multitasks and are programmed to do things in a human way and even more.

Telephones would incorporate Internet connection helping the transmission of files. Reports and letters of meeting in companies can be sent through smartphones. Someone on a mission does no longer need to wait till the end of the mission before submitting the reports, minutes, or whatever administrative writings. Files can be sent to and fro everywhere in limitless time thanks to smartphones. What can't the world do with a mobile phone now?

It is wonderful to notice the extent the world has come to when only a century and a half ago, a man could hardly imagine a world of this kind. Some centuries ago, human beings could not even communicate to their fellows at a mile away if one does not go to them. But today, thanks to smartphones, not only could we communicate with people at millions of miles away but we can also talk to them, feel every single of their breathing, and even see them and their immediate environment instantly as well. Photos, video camera, and games are among the services the smartphones offer.

Technology and science are offering comfort and simplifying the toils in life. Decisions can be made based on data release online. Meetings, conferences, and appointments are held at a distance simultaneously with many people in different places, countries, and on other continents thanks to the applications on smartphones.

Smartphones have upset down the world socially, politically, and even financially [9].

3.2. Smartphones at the core of the social life

Socially speaking, smartphones have become a socializing mean, physical meetings are no longer indispensable, unless for major cases. There is no necessity for young people to go to cyber café to connect to social networks, Facebook, Twitter, WhatsApp, to name but a few. Smartphones are equipped with WIFI apps to connect to local area network (LAN) and wide area network (WAN). Journalists can report at real-time events from the other end of the planet with photos or videos for evidences. The power of the press has increased thanks to smartphones that make photos, shoot videos in secret without the police or the governments' authorities censoring them.

Besides, anybody at all can send information with his/her smartphone online. Therefore, fewer secrets are being held and people are informed and they can mobilize for or against things they like or dislike. The very example of Tunisia and Egypt during the Arab Spring is an illustration [9].

Also, do smartphones help consolidate democracy in developed countries and contribute to build democracy in developing countries. For example, if the police is used to brutalize or prevent people from demonstrating pacifically by beating or firing on the crowd, those in the crowd who have smartphones can shoot video or take picture and share on social networks instantly, then journalists will relay the information on televisions, radios, and so on. This has helped reduce violence on the population in countries where the heads of states are cold-blood dictators.

When the war in Syria started, the government would forbid journalists to tell the truth on what was going on. However, some independent journalists could report the killings perpetrated on the population on social networks anonymously, and this, thanks to their smartphones. Facing all these forces that smartphones offer, they are being needed and used more often.

Therefore, a ruthless war has exploded among telephony companies. Apple, Samsung, Microsoft, and new Chinese companies are fighting to offer a smartphone of the latest generation to their consumers, a strategic tool for the whole world in integrating applications that were used for laptops and desktops before. Moreover, they are struggling to make smartphone tools in every domain of human activities according to the demand.

Table 1 presents a list of some services or applications integrated to smartphones.

Medically, further studies and experimentations are being conducted to associate smartphones to remotely monitor outpatients with incisional wounds and collect postoperative symptom information [10]. Patients can take digital photos with their smartphones and send them to caregivers who are located far away. Likewise, they can also give information on their postoperative symptoms via smartphones. This practice of medicine is known as telemedicine and is expanding nowadays. Telemedicine is remote diagnosis and treatment of various health problems. Therefore, making a follow up of non-hospitalized patients.

Farming robots are lastly created and equipped with a device to send a text message to the farmer after the robot finishes the work bestowed on it. This example shows that smartphones have brought a real revolution in the world and to its every single aspect. Smartphones are no longer tools for bureaucrats only but even on the field where land is cultivated.

Smartphone offers leisure and helps people who get bored during their spare times. Not only do video games, TV channels, radio stations incorporated in it occupy people, but

Applications/services	Description
General	Calculators, alarms, notepads, diaries
Geolocations	Identification and position of devices (ships, stars). GPS techniques are used There is a legal vacuum regarding individuals, so this service has not seen great development.
Sports	Allow real-time values associated with sporting activities (walking, running, and swimming) to be recorded, such as heart rate. Recommends the training to be followed, route exercises.
Medicine	World maps with information on informational epidemic in real time. Application for the control of blood alcohol levels, telerehabilitations, stimulation for disabled people, and control of medicine
Leisure	Search engines for leisure and event venues, music players, videos, films, access to televisions and radio channels, and games
Business	Presentations and video conferences, remote access to applications, online and offline statistics, maps of geographical results, markets analysis, inventory access, presentations of products to clients, and launchings of market campaigns. Direct-to-bill mobile payments, avoiding the used of credit cards for Internet transactions
Social	News of general nature, magazines or newspapers, access to social networks (Twitter, Facebook), and messaging [WhatsApp, Spotbros]
Cloud	Access to files stored in the clouds
Education	Courses, language cooking, translators, books (novels, child education, media, university), and virtual universities

Table 1. Summary of services or applications integrated to smartphones.

smart phones can also help to locate places and venues of events through engine researcher. The advantages that technology is offering to mankind nowadays are unfolding. Things are moving so fast. Many tasks of a man are being suppressed and replaced by technology. We can hardly think of life without technology today to the extent that it seems as if technology can substitute human beings in all their tasks. It is not imaginable to see somebody without telephone for 24 hours today. Telephones are at the core of life and we cannot do without them, not even for a single day. We have got used to their use that we hardly have time to think of the eventual dangers that could occur from them or to think of the negative impact they can generate on a person and the question is: "What are the limits of smartphones?"

4. Behavioral impacts of smartphones

Researches and studies have shown that the extensive use of smartphones can affect people's behavior. Thus, according to Lepp, Barkley, and Karpinski, there is a negative correlation between cell phone use and academic performance [11]. They came out with a possible explanation stipulating that the time spent on the phone or smartphone is missing in academic endeavors.

Besides, students craving for smartphones or cell phones happen to be multitasking [12]. Multitasking robs working memory' capacity on the one hand, and on the other hand, when responding to emotionally gratifying distracters a task is approached in a more superficial way and it takes longer.

Moreover, empiric experiences showed that some people, especially young people, could no more depart from their telephones; not even for a second. They have become addicted to their telephones and probably for having pornographies on them or for entertaining some relationships like making love online, connecting to other people all around the world on social networks or whatever. It takes them the whole day and the whole night. Some can even hardly sleep, eat, or do their assignments as they are students. Therefore, for passing sleepless nights and for not eating, people with excessive and abusive use of smartphones happen to become sick and thin. Moreover, smartphone can reveal to be a source of corporal danger as well.

4.1. A smartphone, a potential cause of crash?

According to the article, "The real reason you're told you put your mobile in flight mode" published on the website Travel Truths, "Some more begrudgingly than others – as their signals interferes with navigation instruments, and could even cause a crash." One thing that is true in this is that any smartphone or a mobile phone has flight-mode application.

In this article, Patrick Smith, a pilot and the author of Cockpit Confidential, answered "yes" to the assumption that smartphone can cause a crash although he added to his answer "technically" and "but it's more an exercise of caution."

He said: "Aircraft electronics are designed and shielded with interference in mind." This statement, though, it mitigates the fear of air crash related to the use of mobile phone on flights, confirms albeit the assumption of a smartphone being a source of explosion to cause fire. He even clearly added that cellular communication can disrupt cockpit equipment but in all likelihood no.

Moreover, France 24 reported on its website that Samsung Galaxy Note 7 was added to no-fly list of some flight companies and the Federal Aviation Agency called on travelers not to fly with this smartphone that is potentially explosive.

Many incidents, a total of 35, were reported relating to the explosion of the said telephone. On September 10, 2016, the phone exploded as a kid was watching a video on it [13]. He was seriously burned and was taken to the hospital; he was afraid since then to touch a smartphone. On September 05, 2016, the explosion of another Samsung Galaxy Note 7 caused damages of an amount of Australian $1800 in a hotel in Australia [14]. The company accepted to refund the amount of money to the hotel. As we said before, there were 35 incidents related to the explosion caused by the Samsung Smartphone.

Samsung Galaxy Note 7 explosion reported by France 24 is not the only incident related to mobile explosion in general.

In Togo, Simlawo et al. reported a case on accident caused by a mobile phone [15]: **"Evisceration caused by the explosion of mobile phone battery: A rare form of domestic accident in a child"**

A 7-year-old child in a second grade was admitted to the Emergency Room of the Regional Hospital Center (RHC) of Lomé Commune for abdominal open trauma with bowel evisceration

http://dx.doi.org/10.5772/intechopen.70260

due to a domestic accident. The child was playing outside the house as the garbage was burning from about 5 m distant. Accidentally, there was an explosion of a mobile phone battery out of the burning garbage. The exploding battery hit the child on the left flank, which led to an abdominal open trauma with bowel evisceration. The bowels could be seen at about 50 cm in evisceration (**Figure 1**). The evisceration was made by an abdominal injury located on the left flank with bruised sides and oval of about 5 cm from the large axle. The injury was not hemorrhagic. The other part of the abdomen had no specific problem. The agent that caused the trauma was brought to the hospital. It was a cubic metallic ball of 5 cm long and 3 cm thick. It was the battery of a mobile phone that exploded while it became overheated. **Figure 2** shows the battery and its exploded portion.

Mobile phones in general use batteries made of lithium ion that is potentially explosive. In the developing countries, there is no particular policy of recycling that can help get rid of these batteries. Although mobile phones are very useful nowadays and almost every household possess one, the management of their damaged batteries must comply with physico-chemical rules. It is also necessary to make a rational and strict management of household wastes in general and in particular, to recycle the mobile phone batteries that are out of use.

4.2. Lack of information on the dangerous aspect of smartphones

Many at times, notice is given on how to use smartphones but hardly a warning is given on the potential danger that can result from a mobile phone. Surprisingly, parents are unaware of dangers faced by children on smartphones [16].

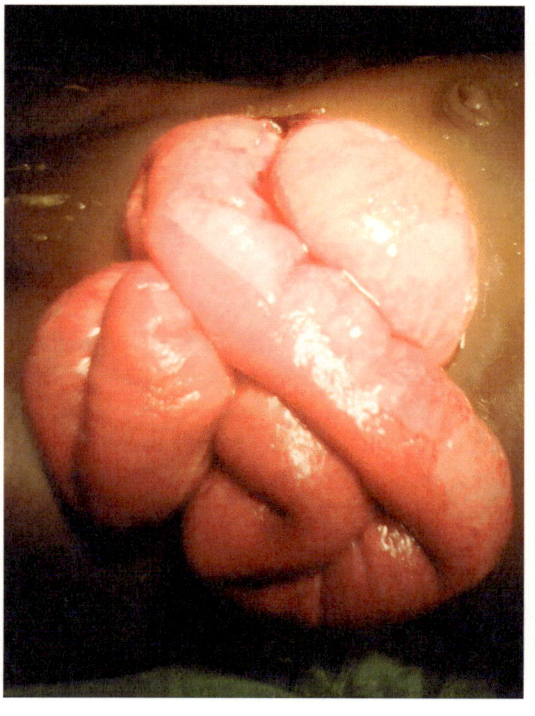

Figure 1. Evisceration of the small intestine on the left edge.

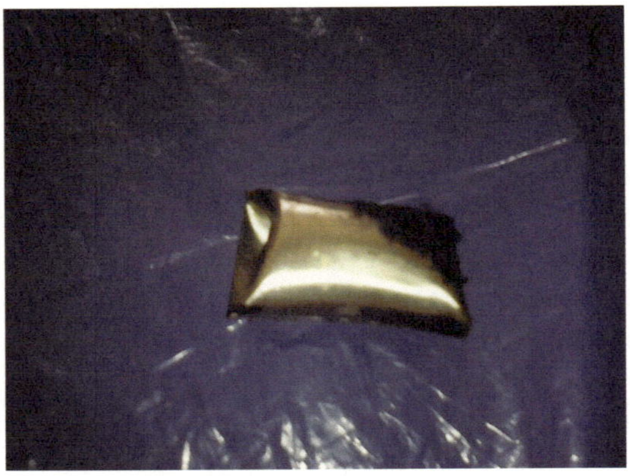

Figure 2. The mobile phone battery modified by the heat.

According to another study on people's opinion about children using a telephone, as the issue is raised by the question, **"Should kids use smartphones?"** 60% say "yes" [17].

With all what that have been said and studied, children's vulnerability vis à vis the use of smartphone is no more to prove. Controversially as we can notice, parents are little or not at all informed on the probable dangers of smartphones and worse again on the protective measures to take. This can't help preserve children's health both psychologically and physically. In the case of the case report on Evisceration due to the mobile phone battery explosion, if information was divulgated on the chemical component of the battery and warning on its potential explosive danger, it could have been avoided to throw the battery in the household garbage. The explosion could be prevented. Therefore, the child could have escaped this accident.

5. Taking measures to limit or eradicate smartphone-related dangers

5.1. Mobile telephony companies

Dispositions should be taken to conceive nonflammable batteries for smartphones, for example, as we suggest working to reduce side effects of smartphones. In a high rate related to incidents that occurred, children are more exposed. Considering applications in smartphones, children and young people are also of high-rate users. Games, video games, cartoons, social networks (Facebook, WhatsApp, twitter) to mention but a few attract and glue this range of population to smartphones. They are targeted by the dangers that may come from smartphones.

5.2. Parents and governments

Twenty percent of parents do not monitor their children's use of smartphones [16]. This is a study conducted in Europe especially in England. It becomes more concerning when survey is made in African countries. No scientific study testifies but when we consider the educational rate of developing and underdeveloped countries in general, it does pass

http://dx.doi.org/10.5772/intechopen.70260

40% (Wikipedia); however, the population rate of the consumers of smartphones passes 70% and of course this rate is dominated by young people and teenagers in average. Therefore, the issue of monitoring children's use of smartphone becomes more delicate. Actually, children master more smartphones than the parents except some parents of course. However, parents should discuss openly on undesirable things their children may come across on Internet. Prepare them on the eventual negative impact it may have on them.

Since parents can't really control everything their children do with smartphones, a protective concept of children protection against smartphone should come from the government as well.

Government should insert in the educational system a program that will inform students and raise awareness on the dangers of smartphones. Government should also ask the companies to conceive smartphones for kids under 18 with restriction on applications. The authorities should control sellers and require from them not to sell smartphones without application restriction under 18.

6. Conclusion

Life has become much more livable with the technology and the creation of mobile phones. Nonetheless, it may affect people's psychology, behavior, and health especially those of children. Awareness should be raised on the dangers of smartphones not only for children but also for adults. Therefore, measures should be taken to protect children and teenagers to ensure their welfare as they use smartphones.

Telephony companies as well as parents should join their effort to figure out an adequate solution for this phenomenon. A mobile phone becomes dangerous when it heats because it can explode. Yet, awareness must be raised on the use of smartphones and mobile phones in general for, as we said before, a telephone has become a real life partner in everything. Therefore, the humanity is in permanent danger.

Author details

Simlawo Kpatékana[1], Boumé Missoki Azanlédji[2]*, Kanassoua Kokou[3], Mihluedo-Agbolan Komlan Anani[2] and Bouame Kokou Tsolanyo[4]

*Address all correspondence to: missokiboume@gmail.com

1 General Surgical Service of RHC of Lomé-Commune, Lome, Togo

2 Pediatric Surgery Service of UHC, Sylvanus Olympio (SO) of Lomé, Lome, Togo

3 Emergent Surgical Service UHC, Sylvanus Olympio (SO) of Lomé, Lome, Togo

4 Galilée Consulting SARL, Lome, Togo

References

[1] Carrey E, Payne KFB, Ahmed N, Goodson A. The benefit of the smartphone in oral and maxillofacial surgery: Smartphone use among maxillofacial surgery trainees and iPhone apps for the maxillofacial surgeon. Journal of Maxillofacial and Oral Surgery. 2015;**14**(2):131-137

[2] Demirci K, Akgönül M, Akpinar A. Relationship of smartphone use severity with sleep quality, depression, and anxiety in university students. Journal of Behavioral Addictions. 2015;**4**(2):85-92

[3] Archbold HA, Guha AR, Shyamsundar S, McBride SJ, Charlwood P, Wray R. The use of multi-media messaging in the referral of musculoskeletal limb injuries to a tertiary trauma unit using: A 1-month evaluation. Injury. 2005;**36**:560-566

[4] Dala-Ali BM, Lloyd MA, Al-Abed Y. The uses of the iPhone for surgeons. Surgeon. 2011;**9**:44-48

[5] Lee J-H, Jung H-K, Lee G-G, Kim H-Y, Park S-G, Woo S-C. Effect of behavioral intervention using smartphone application for preoperative anxiety in pediatric patients. Korean Journal of Anesthesiology. 2013;**65**(6):508-518

[6] Kwon M, Lee JY, Won WY, Park JW, Min JA, Hahn C, et al. Development and validation of a smartphone addiction scale (SAS). PloS One. 2013;**8**(2):e56936

[7] Available from: http://pewinternet.org/Report/2013/Smartphone-Ownership/Pew Research Center, Internet and Technology. [Accessed: April 2017]

[8] Cerdeño E, IT Deputy. Phone evolution and revolution trébol 2013;**65**:17-27

[9] Lotan G, Graef E, Ananny M, Gaffey D, Pearce I, Boyd D. The Arab Spring: The revolutions were tweeted: Information flows during the 2011 Tunisian and Egyptian revolutions. International Journal of Communication. 2011;**5**:1375-1405

[10] Wiseman JT, Fernandes-Taylor S, Barnes ML, Tomsejova A, Scott Saunders R, Craig Kent K. Conceptualizing smartphone use in outpatient wound assessment: Patients' and caregivers' willingness to use technology. Journal of Surgical Research. 2015;**198**(1):245-251

[11] Samaha M, Hawi NS. Relationships among smartphone addiction, stress, academic performance, and satisfaction with life. Computers in Human Behavior. 2016;**57**:321-325

[12] Rosen LD, Whaling K, Carrier LM, Cheever NA, Rokkum J. The media and technology usage and attitudes scale: An empirical investigation. Computers in Human Behavior. 2013;**29**(6):2501-2511

[13] Wolff K. Childhood and Globalization Encyclopedia of Educational Philosophy and Theory. 2016. pp 1-5, Springer Singapore

[14] Available from: http://www.dailymail.co.uk/…/A-Samsung Galaxy Note 7-caused $1800: damages to a Perth Hotel. [Consulted on 14 April 2017, 11 AM]

[15] Simlawo K, Boume AM, Kanassoua K, Sambiani D, Ametitovi E, Gnassingbe K. Evisceration caused by the explosion of mobile phone battery: A rare form of domestic accident in a child. Journal of Pediatric Surgery Case Reports. 2016;**14**:35-37

[16] Available from: http://www.bbc.co.uk/news/technology-26121434/ Parents unware of dangers faced by children on Smartphones-BBC News. [Accessed: April 2017]

[17] Available from: http://www.debate.org /Opinions/ Should-kids-have-smartphones? [Accessed: April 2017]

Chapter 10

Malware Analysis and Detection on Android: The Big Challenge

Abraham Rodríguez-Mota,
Ponciano J. Escamilla-Ambrosio and
Moisés Salinas-Rosales

Additional information is available at the end of the chapter

http://dx.doi.org/10.5772/intechopen.69695

Abstract

The popularization of the use of mobile devices, such as smartphones and tablets, has accelerated in recent years, as these devices have experienced a reduction in cost together with an increase in functionality and services availability. In this context, due to its openness and free availability, Android operating system (OS) has become not only a major stakeholder in the market of mobile devices but has also become an attractive target for cybercriminals. In this chapter, we advocate to present some current trends and results in the Android malware analysis and detection research area. We start by briefly describing the Android's security model, followed by a discussion of the static and dynamic malware analysis techniques in order to provide a general view of the analysis and detection process to the reader. After that, a description of a particular set of software developments, which exemplify some of the discussed techniques, is presented accompanied by a set of practical results. Finally, we draw some conclusions about the future development of the Android malware analysis area. The main contribution of this chapter is a description of the realization of static and dynamic malware analysis techniques and principles that can be automated and mapped to software system tools in order to simplify analyses. Moreover, some details about the use of machine learning algorithms for malware classifications and the use of the hooking software techniques for dynamic analysis execution are provided.

Keywords: malware analysis, android, mobile devices, threat detection, cybersecurity

1. Introduction

Nowadays, mobile devices such as smartphones and tablets have become very popular, due to a reduction in their cost and an increase in their functionalities and services availability.

Moreover, the growing trend of implementing bring your own device (BYOD) policies in organizations has also contributed to the adoption of these technologies, not only for everyday communication activities but to support enterprise systems, industrial applications, and commercial transactions, which raise new security issues. In this scenario, operating systems have also played an important role in the adoption and proliferation of mobile devices and applications, giving also space for the appearance of malicious software (malware). This is the case for the Android OS, which, due to its openness and free availability, has become not only a major stakeholder in the market of mobile devices but has also become an attractive target for cybercriminals.

Google, the Open Handset Alliance manufacturers, and the Android developers' community have made many efforts in order to improve Android's security. However, the emergence and evolution of new security threats continue being an important issue. Therefore, in this chapter, we advocate to present some current trends and results in the Android malware analysis and detection research area. We start by briefly describing the Android's security model, followed by a discussion of some static and dynamic malware analysis techniques in order to provide a general view to the analysis and detection processes to the reader. After that, a description of a particular set of software developments, which exemplify some of the discussed techniques, is presented accompanied by a set of practical results. Finally, a set of conclusions about the future development of the ideas explored in this chapter are drawn.

2. Android security architecture

In a general sense, Android is not only an OS but a platform of three main building blocks: device hardware, Android OS, and the application runtime, see **Figure 1**.

First of all, the Android device hardware block refers to the wide range of hardware configurations where Android can be run, including smartphones, tablets, watches, automobiles, smart TVs, OTT gaming boxes, and set top boxes. Android is processor-agnostic, but it does take advantage of some hardware-specific security capabilities such as ARM eXecute-Never. Secondly, the Android OS building block refers to the Android OS itself, which is built on top of the Linux kernel, thus all device resources are accessed through the operating system. Thirdly, the Android application runtime block refers to the managed runtime used by applications and some system services on Android [2]. In this case, it must be taken into account that applications are written in the Java language and run in the Android runtime (ART). However, many applications, including core Android services and applications, are native applications or included native libraries. Both ART and native applications run with the same security environment, contained with the applications sandbox. Thus, applications get a dedicated part of the file system in which they can write private data, including database and raw files [1].

In this context, in terms of security, Android incorporates industry-leading security features and works with developers and device implementers to keep the Android platform and ecosystem safe. It was designed with multi-layered security that is flexible enough to support an

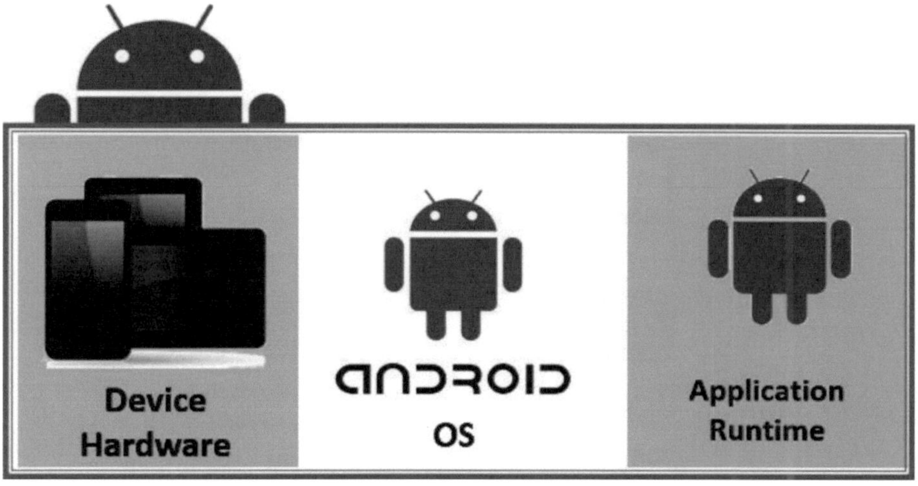

Figure 1. The main Android platform building blocks, adapted from [1].

open platform while still protecting all users of the platform. Security controls were designed to reduce the burden on developers. In this way, security-experienced developers can easily work with and rely on flexible security controls and developers less familiar with security concepts will be protected by safe defaults [1].

Moreover, Android provides a set of key security features, which are: robust security at the OS level through the Linux kernel, mandatory application sandbox for all applications, secure inter-process communication, application signing, and application-defined and user-granted permissions [1]. In the first case, as shown in the Android software stack, see **Figure 2**, each component assumes that the components below are properly secured. In this scheme, with the exception of a small amount of Android OS code running as root, all code above the Linux kernel is restricted by the application sandbox [1]. It is important to notice that the Android kernel is slightly different from a "regular" Linux kernel, the differences are due to a set of features originally added to support Android, and some of them are the low memory killer, wakelocks, anonymous shared memory, alarms, paranoid networking, and Binder. However, Android's security model also takes advantage of the security features offered by the Linux kernel. In a Linux system, which is a multi-user operating system, the kernel can isolate user resources from one another, just as it isolates processes. Consequently, one user cannot access another user's file, unless explicitly granted permission, and each process runs with the identity of the user that started it. In a traditional system, a user ID (UID) is given either to a physical user that can log into the system and execute commands via the shell or to a system service (daemon) that executes in the background. At this point, it is also worth to notice that Android was originally designed without the need for registering different physical users with the system, thus the physical user is implicit and UIDs are used to distinguish applications instead. This forms the basis of Android's application sandboxing [3].

On top of the Linux kernel layer is the hardware abstraction layer (HAL). This layer provides a standard method for creating software hooks between the Android platform stack

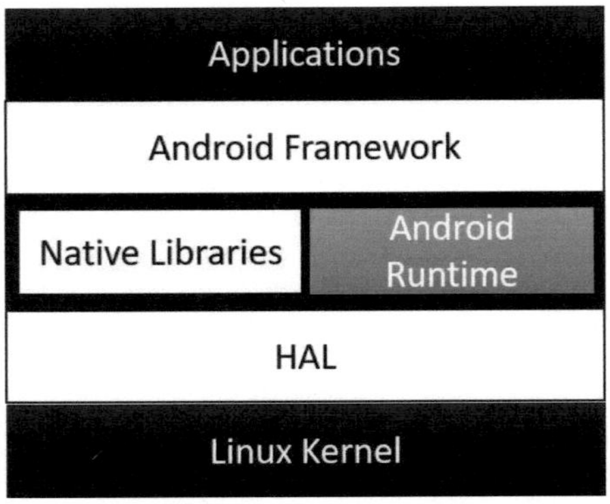

Figure 2. Android software stack, adapted from [1].

and the hardware. The HAL allows to implement functionality without affecting or modifying the higher level system [4]. In the next layer, the libraries component acts as a translation layer between the kernel and the application framework. The native libraries in Android are written in C and C++, most of which are ported from Linux, but are exposed to developers through a Java API. At the same level, there are also components from the Android runtime and core libraries. The virtual machine is an important part of the Android operating system and executes system and third-party applications [4, 5].

At the next level, the Android framework layer provides a suite of services or systems that are useful when writing applications. Commonly referred to as the application programming interface (API) is one of the building blocks for the final system or end-user applications. Finally, at the top most layer, applications component of the Android OS is located, which is the layer closest to the end user. All finished developed products will execute in this space by using the API libraries and the runtime environment [5].

As mentioned before, Android apps are composed of different components and each app is sandboxed by executing it in a separate process with a distinct user ID and assigning it to a private data directory on the file system. The four basic Android application components are *Activities, BroadcastReceivers, ContentProviders,* and *Services.* All components can be interconnected remotely across process boundaries by using different abstractions of Binder inter process communication (IPC) [6]. These interconnections are commonly referred to as inter-component communication and are the primary communication mechanism in Android although it can provide classical channels such as files or sockets. Android apps can either contact system services or communicate directly with each other. User space processes can communicate with each other over Binder IPC via the Binder kernel module. Android's design provides different levels of abstraction for Binder IPC, allowing developers to easily make use of Binder IPC at the application level to connect different apps' components (stubs, proxies, and managers). All inter-component communication (ICC) is built on the top of Binder IPC [7], see **Table 1**.

Moreover, every application that is run on the Android platform must be signed by the developer. Application signing allows developers to identify the author of the application and to update their application without creating complicated interfaces and permissions.

Component	Description
Stubs and proxies	The most basic level of abstraction of Binder IPC. Implement remote procedure calls (RPC) via Binder IPC. A proxy at the caller-side marshals the method parameters into primitive data types and transfers them via IPC to the recipient, where stub unmarshals the primitives into the original parameters and calls the actual method.
System services and managers	Managers are part of the SDK and encapsulate pre-compiled proxies for system apps and services like the location manager service that implement the Android application framework API.
Intents	The highest level of abstraction is so-called intent messages. An intent is a data structure used to provide an abstract description of an operation to be performed by its receiver(s). Common usages of Intents include starting activity components or broadcasting notifications to apps. Since the sender of an intent can both explicitly state the target component and implicitly define potential receivers through a description of the intended action, the actual target app(s) must be resolved at runtime.

Table 1. Inter-component communication (ICC) builds on top of Binder IPC.

Applications that attempt to install without being signed will be rejected by either Google Play or the package installer on the Android device. Application signing ensures that one application cannot access any other application except through well-defined IPC. Applications can be signed by a third-party or self-signed. It is also possible to declare security permissions at the Signature protection level, restricting access only to applications signed with the same key while maintaining distinct UIDs and application sandboxes. A shared application sandbox is allowed via the shared UID feature where two or more applications signed with same developer key can declare a shared UID in their manifest [8].

Android applications can access only their own files and any world-accessible resources on the devices due to the sandboxed nature of Android. However, Android can grant additional, fine-grained access rights to applications in order to allow for richer functionality. Those access rights are called permissions, and they can control access to hardware devices, Internet connectivity, data, or OS services. Applications can request permissions by defining them in the AndroidMAnifest.xml file. At the application install time, Android inspects the list of requested permissions and decides whether to grant them or not. Once granted, permissions cannot be revoked and they are available to the application without any additional confirmation. For some features, explicit user confirmation is required for each accessed object, even if the requesting application has been granted the corresponding permission [3].

Additionally, Android also enforces security by providing preinstalled and user-installed applications. Pre-installed applications work as users applications and as providers of key devices' capabilities that can be accessed by other applications. This application may be a part of the open source Android platform or they may be developed by a device manufacturer for a specific device. On the other hand, Google Play, Android's application official store, offers users hundreds of thousands of applications, including many third-party applications [1].

Outside these security features, Android also provides a set of cloud-based services that are available to compatible Android devices with Google Mobile Services. These services are not part of the Android Open Source Project, but are included on many devices. See **Figure 3**.

Figure 3. The primary Google security services.

Briefly described, *Google Play* is a collection of services that allow users to discover, install, and purchase applications from their Android device or the web. It also provides community review, application license verification, application security scanning, and other security services. The *Android update service* delivers new capabilities and security updates to selected Android devices, including updates through the web or over the air (OTA). The *Application services* term refers to a set of frameworks that allow Android applications to use cloud capabilities such as (backing up) application data and settings and cloud-to-device messaging (C2DM) for push messaging. The *Verify Apps* service warns or automatically blocks the installation of harmful applications, and continually scan applications on the device, warning about or removing harmful apps. *SafetyNet* is a privacy preserving intrusion detection system to assists Google tracking and mitigating known security threats in addition to identify new security threats. The *SafetyNet Attestation* is a third-party API to determine whether a device is CTS compatible. Attestation can also assist to identify the Android application communicating with the application server. Finally, the Android device manager is a Web and Android application to locate lost or stolen devices [1].

As it can be observed from the previous description, Android has become a continuously evolving complex ecosystem composed of multiple subsystems and services that put together an enormous challenge in terms of security. In this context, in the following section, a brief discussion of some attempts to conceptualize and characterize the Android attack surface and key security challenges is presented prior to the later discussion of some of the main malware analysis and detection techniques, as an initial landmark from where techniques and research approaches presented later on may be better referred to or mapped to specific security aspects of the Android ecosystem.

http://dx.doi.org/10.5772/intechopen.69695

3. The android attack surface

An attack surface is a term used to identify the characteristics of a target that makes it vulnerable to attack. An attack vector generally refers to the means by which an attacker performs an attack. In other words, an attack surface refers to the code that an attacker can execute and therefore can attack. In contrast to an attack vector, an attack surface does not depend on the attackers' actions or require a vulnerability to be present, it describes where in code vulnerabilities might be waiting to be discovered. Generally, the size of a target's attack surface is directly proportional to how much it interfaces with other system. Similar to attack vectors, attack surfaces can be discussed both in general and in increasingly specific terms. It is a common result that by studying one particular attack surface, additional attack surfaces are revealed [9].

By focusing on particular risky attack surfaces, a system can be attacked or secured more quickly. Several properties are important when identifying attack surfaces, some of them are: attack vectors, privileged gained, memory safety, and complexity. Because Android devices have such a large and complex set of attack surfaces, it is necessary to divide them [9]. **Figure 4** exemplifies some of the more general attack surfaces for Android devices together with some attack vectors and propagation mechanisms.

The remote attack surface is the largest and most attractive attack surface exposed by an Android device. This name, which is also an attack vector classification, aims to express the fact that the attacker does not need to be physically located near the victim. Instead, attacks are executed over a computer network, usually the Internet. Various properties further divide this surface into distinct groups, see **Figure 4**. The Remote attack surface address the various attack surfaces exposed to code that is already executing on a device. The privileges required to access these attack surfaces vary depending on how the various endpoints are secured. When an attacker has achieved arbitrary code execution on a device, the next logical step is to escalate privileges, either in the kernel space or under the root or system user. The physical attack surfaces give name to the attacks that require physically touching a device, in contrast to physical adjacency where the attacker only needs to be within a certain range of the target. Third-party modification attack surface relates to attack surfaces associated to the modification of various parts of an Android device system, as many parties involved in creating Android devices tend to make extensive changes as a part of their integration process [9].

Unfortunately, on the top of this complexity, Android's security analysis also requires to take into account a set of Android's security challenges such as: fragmentation, malware, management tool selection, user behavior, and compartmentalization [10]. Fragmentation challenge refers to the complexity associated to the wide range of Android-modified versions implemented on different devices. Malware challenge advocates to the rapid increase of malicious applications development and sophistication targeting the Android OS. Management tool selection challenge relates to the selection of management tools, which can avoid overlapping or conflicting features, as well as to maximize IT productivity. The user behavior challenge refers to the need for encouraging users to comply with good security policies and practices. Lastly, compartmentalization describes the challenge of providing dual personal and mobile virtualization, which separates a single device into different personal environments [11].

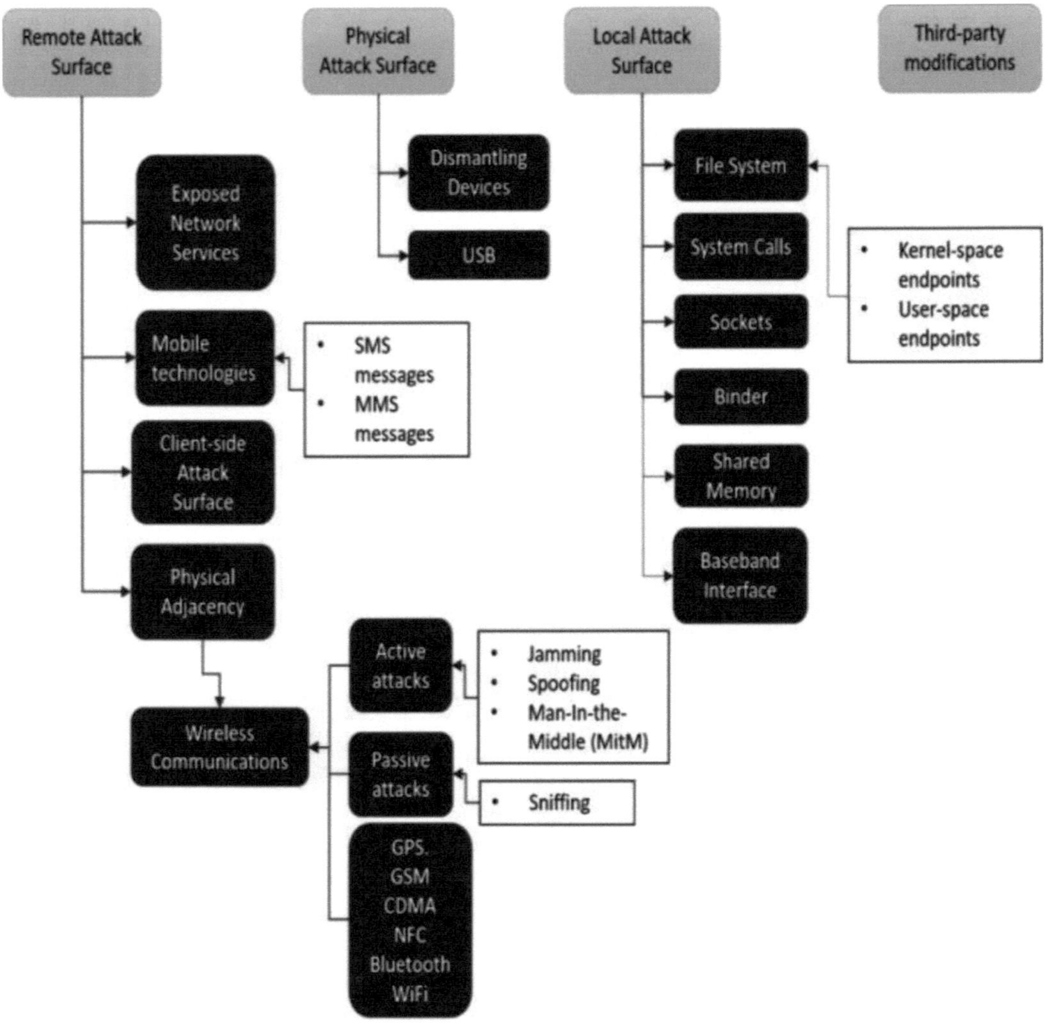

Figure 4. Android's main attack surfaces, based on descriptions in [9].

4. Android malware

Android malware can be characterized in different ways: in [12], a systematic characterization is proposed ranging from their installation, activation, to the carried malicious payloads. Thus, malware installation can be generalized into three main social engineering-based techniques: repackaging, update attack, and drive-by download. Repackaging is one of the most common techniques that malware authors use to piggyback malicious payloads into applications. In essence, malware authors get an application file, disassemble them, enclose malicious payloads, reassemble, and submit the new application to an official or alternative market. Users could be vulnerable by being enticed to download and install these infected applications. In the case of the update attack, instead of enclosing the payload as a whole only an update component is included, which will fetch or download the malicious payloads at runtime. Because the malicious payload is in the "updated" application, not the original application itself, it is stealthier than the malware installation technique that directly includes the entire malicious payload in the first place. The third technique applies the traditional drive-by download attack to mobile space. Though they are not directly exploiting mobile browser vulnerabilities, they are essentially enticing users to download "interesting" or "feature-rich" applications. This is only a set of common techniques, other threats include combinations of the previous techniques, as well as approaches such as "spyware," which intend to be installed to victim's phones on purpose; fake apps that masquerade as the legitimate applications but stealthily perform malicious actions, such as stealing users' credential; applications that provide the functionality they claimed, they are not fake ones, but that intentionally include malicious functionality, which is unknown to users. At last, a group of applications that rely on the root privilege to function well. The leverage known root exploits to escape from the built-in security sandbox [12].

5. Trends of android malware detection

Malware detection as a discipline combines multiple techniques and principles; Zaki Mas'ud et al. [13] have proposed a general classification including four main categories, see **Figure 5**.

Detection techniques can be classified into three detection techniques: signature-based (SB), anomaly-based (AB), and specification-based (SPB) detection. Signature-based detection refers to the malware detection by comparing the application signature or pattern captured with a database of known attacks or threats. AB detection monitors regular activities in the devices and looks for any behavior that deviates from the normal pattern. Similar to AB detection, SPB detection also monitors for any deviation but rather than detecting the occurrence of specific attack patterns; it monitors for deviation of their behavior from the normal specification. The detection analysis category involves reverser engineering techniques aimed to obtain information about the behavior of a malware in its environment. On the one hand, in static analysis, detection is done through the source code, binary, or the API level without the execution of the Android malware. On the other hand, dynamic detection detects malware through the execution behavior of the malware. In this case, the detection is done through monitoring the execution of Android malware activity at runtime. The detection deployment platform category helps to identify whether the malware detection is deployed in the host or on a remote server. In host detection, all the activity of the device is monitored, analyzed, and processed in the device itself. Meanwhile remote deployment requires a remote server, which monitors the activity of the device on the device but performs the analysis and detection process on the remote server. Another important category is the audit data source monitored in the detection process. The data source collected in the Android malware detection can be traced within the five Android framework layers (i.e. application, application framework, Android runtime, libraries, and Linux kernel layers). In addition, network traffic data can also be monitored for any malicious communication activity through the network [13]. Multiple

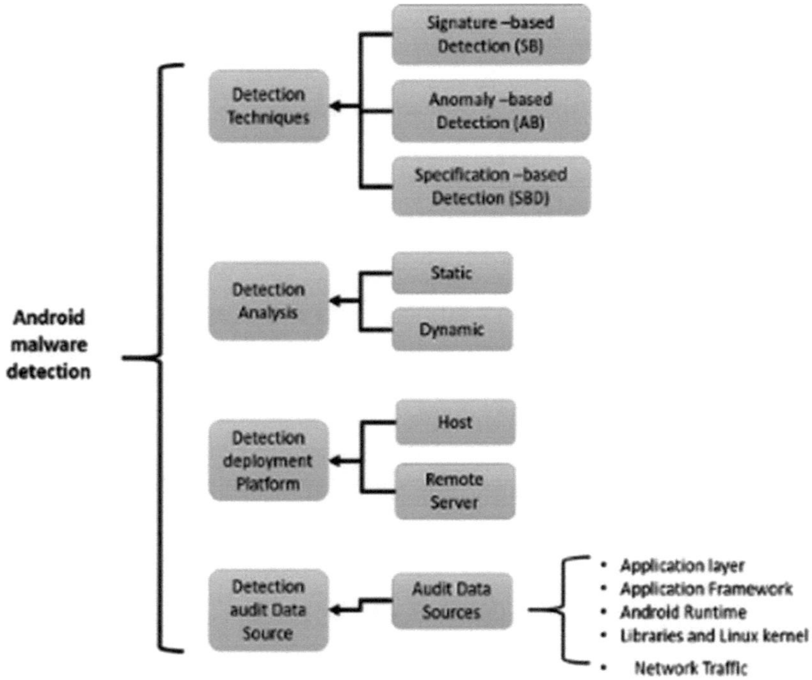

Figure 5. Classification of Android malware detection approaches.

researchers have analyzed different approaches; **Figures 6** and **7** provide an overview, based on the descriptions presented in [14], of different features and algorithms utilized for static and dynamic malware analysis in different research works.

Nowadays, most detection techniques for Android malware use statically extracted data from the AndroidManifest.xml file or Android API function calls, as well as dynamically obtained information from network traffic and system call tracing [15]. Moreover, most current detection systems equipped with a database of regular expressions that specify byte or instruction sequences that are considered malicious are largely based on syntactic signatures and employ static analysis techniques. Unfortunately, static and signature-based analysis techniques can be evaded by malware applications using techniques, such as polymorphism, metamorphism, and dynamic code loading [16].

Dynamic analysis defines a set of rules for the application behavior, which are challenged for an application according to a possible attack. An event is simulated for each rule and the triggered behaviors are checked to detect malware applications. In some cases, modern malicious applications are capable to evade dynamic analysis as they become aware of the analysis environment, or due to the inability of the malware sample to obtain some required external data or service [16].

As security threats evolve, static and dynamic analysis techniques are less capable to identify malware code by their own. Thus, hybrid approaches combine aspects of both static and dynamic

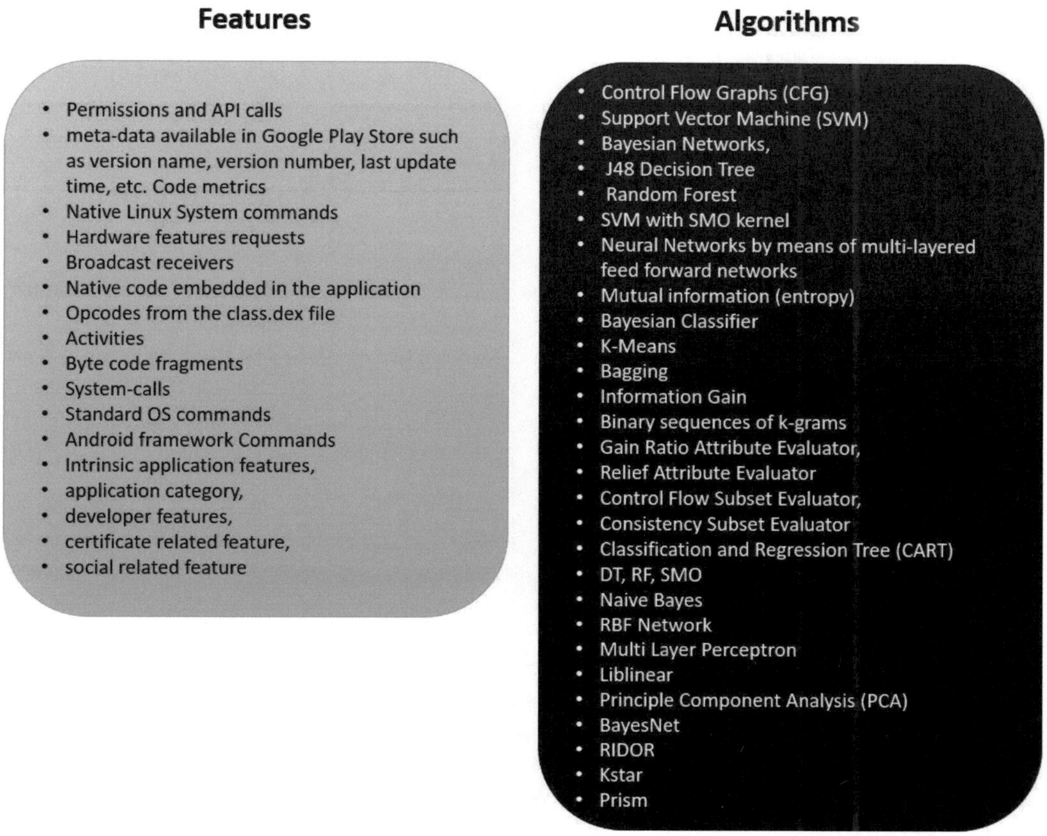

Figure 6. Some common static analysis features and algorithms that are used to process them for different research approaches, based on [14].

analysis [17]. The implementation of hybrid solutions for malware analysis and detection is not a new approach in the PC anti-malware literature [18]. However, the particular characteristics and constraints of mobile devices have defined a new area for their own. In this sense, for example, even when malware analysis and detection schemes can be deployed either on a local basis or offloaded to an external equipment, like a remote server, differences between the mobile and PC ecosystems imply a totally different approach to solve this challenge in both cases. In the particular case of mobiles devices, most current client side security solutions include anti-virus or anti-malware applications installed on the devices to protect them against known applications installed on the mobile devices based on known signatures of malicious applications [19]. On the other hand, cloud-based systems are mainly designed to offload a significant part of the operation to the cloud. Both approaches entail performance constraints and disadvantages. As an example, in applications installed on mobile devices aiming to provide real-time protection, there is an associated decrement in the device's performance and battery life, while cloud-based approaches making use of high end resources cannot offer real-time protection by their own, as they can leave devices vulnerable when connectivity with the server is poor [20].

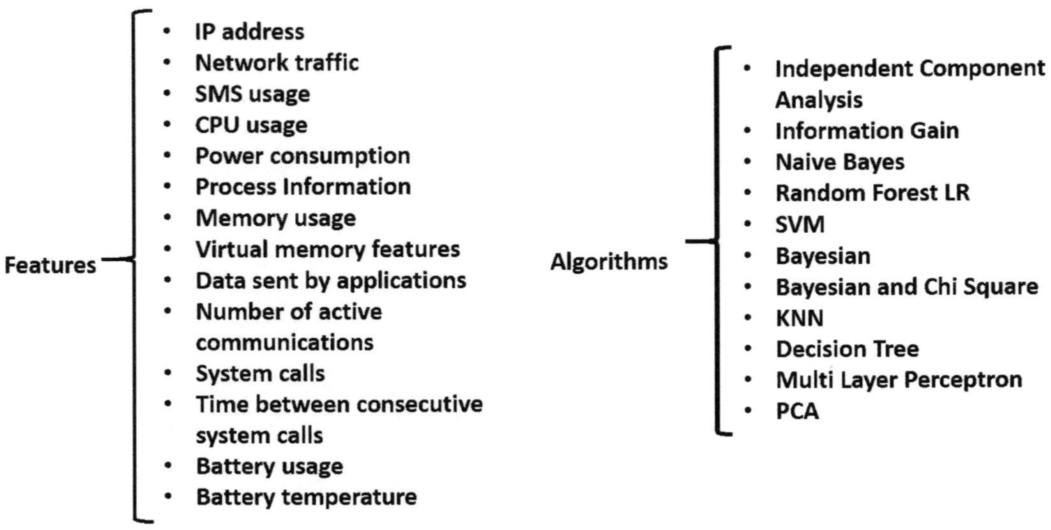

Figure 7. Some common dynamic analysis features and algorithms that are used to process them for different research approaches, based on [14].

Unlike hybrid detection and analysis schemes taking advantage of both static and dynamic analysis, as well as from local and remote combined implementation or execution, are generally common for PC equipment, these schemes are not common for mobile devices. Most solutions combine static and dynamic analysis methods or local and remote deployments but not both of them, as this would require too many compromises to be achieved with the current technologies [18].

6. Conclusions

The continuous development and fast change of the smart devices market has promoted an increase in the number of services and applications offered. As these devices integrate to the users every day activities, they become very attractive targets for cyber criminals. In this sense, malicious software (malware) has become a main security issue in this area. Although malware is not a new problem in the IT industry, differences between PC and smart devices make smart devices security a different problem bounded to the particular features of mobile devices. Moreover, the big number of stakeholders ranging from device manufactures to communication service providers creates a highly heterogeneous environment where attack surfaces characterization becomes a very complex task. In this context, this chapter aimed to present an overview of the fundamental aspects for Android malware analysis and detection.

As it can be deduced from the information discussed above, generally speaking there is a core set of analysis techniques and resource data that have been utilized in multiple research approaches in order to identify and detect malicious software. This observation may be obvious as the identified features are core elements of the Android security architecture, although there is not full agreement in the best technique or procedures for effective malware detection. It is important to notice that machine learning has an important role in most of the discussed approaches and in the state of the art for malware analysis, and in some cases, reported results look highly promising, but there is always the problem of having a limited number of samples to test all possible threats. Additionally, with the current vast set of analysis and reverse engineering set of tools, which are implemented under different technologies and analysis approaches, integration seems a very difficult task to achieve. Moreover, different tools provide multiple and different levels of automation. However, a need for automating most of the process is still an important issue as most of the analysis in the identification of new threats continues to be a human task.

Finally, it is expected that the information presented in this chapter would help readers to obtain a general view of the Android malware analysis and detection area from where she or he can visualize new avenues of research.

Acknowledgements

The authors acknowledge the support from the Mexican Consejo Nacional de Ciencia y Tecnología (CONACYT) under grant number 216747 and Instituto Politécnico Nacional-Secretaria de Investigación y Posgrado under grant numbers 1894 and 20170344.

Author details

Abraham Rodríguez-Mota[1,3]*, Ponciano J. Escamilla-Ambrosio[2] and Moisés Salinas-Rosales[2]

*Address all correspondence to: arodrigm@cic.ipn.mx

1 ESIME Zacatenco, Instituto Politécnico Nacional, Mexico City, Mexico

2 CIC, Instituto Politécnico Nacional, Mexico City, Mexico

3 Centro de Investigación en Computación, Laboratorio de Ciberseguridad, Mexico City, Mexico

References

[1] Android. Security [Internet]. Available from: https://source.android.com/security [Accessed: 12-12-2016]

[2] Android. ART and Dalvik [Internet]. Available from: https://source.android.com/devices/tech/dalvik/ [Accessed: 02-10-2017]

[3] Elenkov N. Android Security Internals. An in-Depth Guide to Android's Security Architecture. USA: No Starch Press; 2015 401 p

[4] Android. Android Interfaces and Architecture [Internet]. Available from: https://source.android.com/devices/ [Accessed: 20-02-2017]

[5] Gunasekera SA. Android Apps Security. New York: Apress; 2012 2021 p

[6] Backes M, Bugiel S, Gerling S. Scippa: System-centric IPC provenance on Android. In: 30th Annual Computer Security Applications Conference; December 08-12-2014; New Orleans, Louisiana, ACM, New York, NY, USA; 2014. pp. 36-45. DOI: http://dx.doi.org/10.1145/2664243.2664264

[7] Enck W, Ongtang M, McDaniel P. Understanding android security. IEEE Security & Privacy. 2009;**7**(1):50-57. DOI: 10.1109/MSP.2009.26

[8] Android. Application Signing [Internet]. Available from: https://source.android.com/security/apksigning/index.html [Accessed: 03-05-2017]

[9] Drake JJ, Fora PO, Lainer Z, Mulliner C, Ridley SA, Wicherski G. Android Hacker's Handbook. USA: Wiley; 2014

[10] Mathias C./TechTarget. Top five Android device management security challenges [Internet]. Available from: http://searchmobilecomputing.techtarget.com/tip/Top-five-Android-device-management-security-challenges [Accessed: 02-02-2016]

[11] Rodríguez Mota A, Escamilla Ambrosio PJ, Aguirre Anaya E, Acosta Bermejo R, Villa-Vargas LA. Improving Android mobile application development by dissecting malware analysis data. In: 4th International Conference in Software Engineering Research and Innovation; 27-29 April 2016; Puebla, Mexico. IEEE; 2016. pp. 1-6. DOI: 10.1109/CONISOFT.2016.21

[12] Zhou Y, Jiang X. Dissecting Android malware: Characterization and evolution. In: 2012 IEEE Symposium on Security and Privacy. 2012. DOI: 10.1109/SP.2012.16

[13] Zaki Mas'ud M, Dahib S, Abdollah MF, Selamat SR, Yusof R. Android malware detection system classification. Research Journal of Information Technology. 2014;**6**:325-341. DOI: 10.3923/rjit.2014.325.341

[14] Baskaran B, Ralescu A. A study of android malware detection techniques and machine learning. In: Phung PH, Shen J, Glass M, editors. Modern Artificial Intelligence and Cognitive Science; 22-23 April 2016. Dayton, OH, USA: CEUR; 2016. p. 15-23

[15] Afonso VM, Favero de Amorim M, Grégio ARA, Barroso Junquera G, Lício de Geus P. Identifying android malware using dynamically obtained features. Journal of Computer Virology and Hacking Techniques. 2015;**11**(1):9-17. DOI: 10.1007/s11416-014-0226-7

[16] Moser A, Kruegel C, Kirda E. Limits of static analysis for malware detection. In: Twenty Third Annual Conference in Computer Security Applications; 12 2007; Miami Beach, FL, USA. IEEE; 2007. DOI: 10.1109/ACSAC.2007.21

[17] Damodaran A, Di Troia F, Vissagio CA, Austin TH, Stamp M. A comparison of static, dynamic, and hybrid analysis for malware detection. Journal of Computer Virology and Hacking Techniques. 2017;**13**(1):1-12. DOI: 10.1007/s11416-015-0261-z

[18] Rodríguez Mota A, Escamilla Ambrosio PJ, Morales Ortega S, Salinas Rosales M, Aguirre Anaya E. Towards a 2-hybrid Android malware detection test framework. In: International Conference on Electronics, Communications and Computers (CONIELECOMP); Cholula, Puebla, México. 2016. pp. 54-61. DOI: 10.1109/CONIELECOMP.2016.7438552

[19] Penning N, Hoffman M, Nikolai J, Wang Y. Mobile malware security challenges and cloud-based detection. In: International Conference on Collaboration Technologies and Systems (CTS); Minneapolis, Minnesota. IEEE; 2014. pp. 181-188. DOI: 10.1109/CTS.2014.6867562

[20] Damopoulos D, Kambourakis G, Portokalidis G. The best of both worlds: A framework for the synergistic operation of host and cloud anomaly-based IDS for smartphones. Seventh European Workshop on System Security (EuroSec'14). 2014;**6**:1-6:6. DOI: 10.1145/2592791.2592797